PERSONALIZATION TECHNIQUES AND RECOMMENDER SYSTEMS

SERIES IN MACHINE PERCEPTION AND ARTIFICIAL INTELLIGENCE*

Editors: **H. Bunke** (Univ. Bern, Switzerland)
P. S. P. Wang (Northeastern Univ., USA)

Vol. 55: Web Document Analysis: Challenges and Opportunities
(Eds. *A. Antonacopoulos and J. Hu*)

Vol. 56: Artificial Intelligence Methods in Software Testing
(Eds. *M. Last, A. Kandel and H. Bunke*)

Vol. 57: Data Mining in Time Series Databases y
(Eds. *M. Last, A. Kandel and H. Bunke*)

Vol. 58: Computational Web Intelligence: Intelligent Technology for Web Applications
(Eds. *Y. Zhang, A. Kandel, T. Y. Lin and Y. Yao*)

Vol. 59: Fuzzy Neural Network Theory and Application
(*P. Liu and H. Li*)

Vol. 60: Robust Range Image Registration Using Genetic Algorithms and the Surface Interpenetration Measure
(*L. Silva, O. R. P. Bellon and K. L. Boyer*)

Vol. 61: Decomposition Methodology for Knowledge Discovery and Data Mining: Theory and Applications
(*O. Maimon and L. Rokach*)

Vol. 62: Graph-Theoretic Techniques for Web Content Mining
(*A. Schenker, H. Bunke, M. Last and A. Kandel*)

Vol. 63: Computational Intelligence in Software Quality Assurance
(*S. Dick and A. Kandel*)

Vol. 64: The Dissimilarity Representation for Pattern Recognition: Foundations and Applications
(*Elżbieta Pękalska and Robert P. W. Duin*)

Vol. 65: Fighting Terror in Cyberspace
(*Eds. M. Last and A. Kandel*)

Vol. 66: Formal Models, Languages and Applications
(Eds. *K. G. Subramanian, K. Rangarajan and M. Mukund*)

Vol. 67: Image Pattern Recognition: Synthesis and Analysis in Biometrics
(Eds. *S. N. Yanushkevich, P. S. P. Wang, M. L. Gavrilova and S. N. Srihari*)

Vol. 68 Bridging the Gap Between Graph Edit Distance and Kernel Machines
(*M. Neuhaus and H. Bunke*)

Vol. 69 Data Mining with Decision Trees: Theory and Applications
(*L. Rokach and O. Maimon*)

Vol. 70 Personalization Techniques and Recommender Systems
(Eds. *G. Uchyigit and M. Ma*)

---

*For the complete list of titles in this series, please write to the Publisher.

Series in Machine Perception and Artificial Intelligence – Vol. 70

PERSONALIZATION TECHNIQUES AND RECOMMENDER SYSTEMS

Editors

Gulden Uchyigit
Imperial College, UK

Matthew Y. Ma
Scientific Works, USA

World Scientific

NEW JERSEY • LONDON • SINGAPORE • BEIJING • SHANGHAI • HONG KONG • TAIPEI • CHENNAI

Published by
World Scientific Publishing Co. Pte. Ltd.
5 Toh Tuck Link, Singapore 596224
USA office: 27 Warren Street, Suite 401-402, Hackensack, NJ 07601
UK office: 57 Shelton Street, Covent Garden, London WC2H 9HE

British Library Cataloguing-in-Publication Data
A catalogue record for this book is available from the British Library.

PERSONALIZATION TECHNIQUES AND RECOMMENDER SYSTEMS
Series in Machine Perception and Artificial Intelligence — Vol. 70

Copyright © 2008 by World Scientific Publishing Co. Pte. Ltd.

All rights reserved. This book, or parts thereof, may not be reproduced in any form or by any means, electronic or mechanical, including photocopying, recording or any information storage and retrieval system now known or to be invented, without written permission from the Publisher.

For photocopying of material in this volume, please pay a copying fee through the Copyright Clearance Center, Inc., 222 Rosewood Drive, Danvers, MA 01923, USA. In this case permission to photocopy is not required from the publisher.

ISBN-13 978-981-279-701-8
ISBN-10 981-279-701-7

Printed in Singapore.

Preface

The phenomenal growth of the Internet has resulted in the availability of huge amounts of online information, a situation that is overwhelming to the end-user. To overcome this problem personalization technologies have been extensively employed across several domains to provide assistance in filtering, sorting, classifying and sharing of online information.

The objective of this book is to foster the interdisciplinary discussions and research in the diversity of personalization and recommendation techniques. These techniques depend on various sources such as domain knowledge, user modeling and user demographics. These fields of research are now being covered by several cross disciplinary societies. It is also the goal of this book to foster the discussions between researchers in pattern recognition community and those in other societies, and address personalization techniques at a broader level.

The first International Workshop on Web Personalization, Recommender Systems and Intelligent User Interfaces (WPRSIUI'05) was organized to address issues related to user interfaces, personalization techniques and recommender systems. It was held in Reading, UK in October 2005. The program committee consisted of a group of well-known researchers and practitioners in the area. Twenty papers were presented at the workshop, the topics ranging from user modeling, and machine learning, to intelligent user interfaces and recommender systems. To solicit papers for this book, authors of the best papers from the workshop were invited to resubmit their extended versions for review along with other papers submitted through the open call. After a prestigious selection process involving two rounds of committee reviewing followed by editors' final review, we are delighted to present the following twelve (12) papers.

The first paper "Personalization-Privacy Tradeoffs in Adaptive Information Access" is an invited contribution by Prof. Barry Smyth. This paper presents the challenges of adapting different devices such as mobile phones to access online information.

The next three papers discuss issues related to user modeling techniques. In "A Deep Evaluation of Two Cognitive User Models for Personalized Search ", Fabio Gasparetti and Alessandro Micarelli present a new technique for user modelling which implicitly models the user's preferences. In "Unobtrusive User Modeling for Adaptive Hypermedia", Hilary Holz, Katja Hofmann and Catherine Reed present a user modeling technique which implicitly models the user's preferences in an educational adaptive hypermedia system. In "User Modelling Sharing for Adaptive e-Learning and Intelligent Help", Katerina Kabassi, Maria Virvou and George Tsihrintzis present a user modeling server with reasoning capability based on multicriteria decision making theory.

Continuing on from the user modeling theme the next three papers discuss issues related to collaborative filtering. In "Experimental Analysis of Design Choices in Multi-Attribute Utility Collaborative Filtering on a Synthetic Data Set", Nikos Manouselis and Constantina Costopoulou present the experimental analysis of several design options for three proposed multiattribute utility collaborative filtering algorithms. In "Efficient Collaborative Filtering in Content-Addressable Spaces", Shlomo Berkovsky, Yaniv Eytani and Larry Manevitz describe a fast heuristic variant of a collaborative filtering algorithm that decreases the computational effort required by the similarity computation and neighbourhood formation stages. In "Identifying and Analyzing User Model Information from Collaborative Filtering Datasets ", Josephine Griffith, Colm O'Riordan and Humphrey Sorensen present a technique of extracting features from the collaborative filtering datasets to be used in modeling groups of users.

Finally the last five papers discuss issues related to content-based recommender systems, hybird systems and machine learning methods. In "Personalization and Semantic Reasoning in Advanced Recommender Systems", Yolanda Blanco Fernandez, Jose Pazos Arias, Alberto Gil Solla, Manuel Ramos Cabrer and Martin Lopez Nores present a hybrid-based recommender system framework which uses semantic information for user modeling. In "Content Classification and Recommendation Techniques for Viewing Electronic Programming Guide on a Portable Device", Jingbo Zhu, Matthew Ma, Jinghong Guo and Zhenxing Wang present a content-based recommender system which presents a personalized browsing and recommendations of TV programs. In "User Acceptance of Knowledge-based Recommenders", Alexander Felfering, Eric Teppan and Bartosz Gula present a knowledge based recommender system for e-commerce. In "Restricted Random Walks for Library Recommendations", Markus Franke and Andreas

Geyer-Schulz present an implicit recommender system which uses restricted random walks for a library application system. In "An Experimental Study of Feature Selection Methods for Text Classification", Gulden Uchyigit and Keith Clark present a comparative study of feature selection method. The above twelve papers represent many interesting research efforts and cover several main categories of personalization and recommendation. This book is dedicated to bringing together recent advancements of personalization techniques for recommender systems and user interfaces. It is also of particular interest to researchers in industry intending to deploy advanced techniques in their systems.

Acknowledgment

The editors would like to acknowledge the contribution and support from all authors in this book and many of invaluable comments from our reviewers including the program committee of the first International Workshop on Web Personalization, Recommender Systems and Intelligent User Interfaces (WPRSIUI'05). They are: Liliana Ardissono, Marko Balabanovic, Chumki Basu, Robin Burke, Joaquin Delagado, Jinhong K. Guo, Xiaoyi Jiang, Mark Levene, Sofus Macskassy, Dunja Mladenic, Ian Soboroff, David Taniar, Patrick Wang and Jingbo Zhu.

Finally, we would like to express our gratitude to Prof. X. Jiang and Prof. P.S.P. Wang, the editors-in-chief of International Journal of Pattern Recognition and Artificial Intelligence (IJPRAI).

G. Uchyigit and M. Ma

Contents

Preface v

User Modeling and Profiling 1

1. Personalization-Privacy Tradeoffs in Adaptive Information Access 3

   B. Smyth

2. A Deep Evaluation of Two Cognitive User Models for Personalized Search 33

   F. Gasparetti and A. Micarelli

3. Unobtrusive User Modeling For Adaptive Hypermedia 61

   H. J. Holz, K. Hofmann and C. Reed

4. User Modelling Sharing for Adaptive e-Learning and Intelligent Help 85

   K. Kabassi, M. Virvou and G. A. Tsihrintzis

Collaborative Filtering 109

5. Experimental Analysis of Multiattribute Utility Collaborative Filtering on a Synthetic Data Set 111

   N. Manouselis and C. Costopoulou

6. Efficient Collaborative Filtering in Content-Addressable Spaces   135

    S. Berkovsky, Y. Eytani and L. Manevitz

7. Identifying and Analyzing User Model Information from Collaborative Filtering Datasets   165

    J. Griffith, C. O'Riordan and H. Sorensen

Content-based Systems, Hybrid Systems and Machine Learning Methods   189

8. Personalization Strategies and Semantic Reasoning: Working in tandem in Advanced Recommender Systems   191

    Y. Blanco-Fernández et al.

9. Content Classification and Recommendation Techniques for Viewing Electronic Programming Guide on a Portable Device   223

    J. Zhu, M. Y. Ma, J. K. Guo and Z. Wang

10. User Acceptance of Knowledge-based Recommenders   249

    Alexander Felfernig and Erich Teppan[1]

    A. Felfernig, E. Teppan and B. Gula

11. Using Restricted Random Walks for Library Recommendations and Knowledge Space Exploration   277

    M. Franke and A. Geyer-Schulz

12. An Experimental Study of Feature Selection Methods for Text Classification   303

    G. Uchyigit and K. Clark

Subject Index   321

---

[1] felfernig@uni-klu.ac.at, teppan@uni-klu.ac.at

PART 1
User Modeling and Profiling

Chapter 1

Personalization-Privacy Tradeoffs in Adaptive Information Access

Barry Smyth

Adaptive Information Cluster
School of Computer Science and Informatics
University College Dublin
Belfield, Dublin 4, Ireland
Barry.Smyth@ucd.ie

As online information continues to grow at an exponential rate our ability to access this information effectively does not, and users are often frustrated by how difficult it is to locate the right information quickly and easily. So-called personalization technology is a potential solution to this information overload problem: by automatically learning about the needs and preferences of users, personalized information access solutions have the potential to offer users a more proactive and intelligent form of information access that is sensitive to their long-term preferences and current needs. In this paper, we document two case-studies of the use of personalization techniques to support information browsing and search. In addition, we consider the inevitable privacy issues that go hand-in-hand with profiling and personalization techniques and highlight the importance of striking the right balance between privacy and personalization when it comes to the development and deployment of practical systems.

1.1. Introduction

The success of the information revolution has been largely characterized by the incredible growth in the information that is available, as the Internet and electronic media continues to expand at an incredible pace. There is a darker side to this revolution, however, as users are becoming increasingly frustrated by how difficult it can be to access the right information at the right time. This is exacerbated by a number of factors beyond the obvious issues such as the sheer quantity and diversity of information that is available. For example, we are using a wide range of devices to access online information, from the traditional PC to the mobile phone, PDAs, and Interactive TV (ITV). These various devices have a wide range of limitations that present their own challenges when it comes to adapting services for their capabilities. In addition, the type of person regularly using information services has changed over the years. Today the vast majority of users are not the information experts anticipated by many information access technologies in the past. For example, the average Web searcher is not an information retrieval expert and cannot readily produce the sort of meaningful and informative queries that most search engines require in order to efficiently respond to a user's needs.

In response to these challenges researchers have highlighted the potential value of so-called *personalized information services*,[2,10,22,24,26,28,30–32] services that are capable of adapting to the learned needs and priorities of individual and groups of users. Personalization technology combines ideas from user profiling, information retrieval, artificial intelligence and user interface design, in order to develop information services that are more proactive and sensitive to our individual preferences. By monitoring our online behavior, for example, it is possible to construct a user profile that accurately captures our information preferences and this profile can then be used to adapt a particular information service in response to these preferences.

This paper will describe two case-studies for personalizing two very different modes of information access (browsing versus search) in two very different content domains (the Mobile Internet versus the World-Wide Web). To begin with, we will focus on the Mobile Internet where *information browsing* is the primary mode of information access and we will describe how personalization techniques can be used to automatically adapt the structure of a mobile portal to reflect the preferences of an *individual* user. In the second case-study we will focus on the more traditional Web, but this time examine how personalization techniques can be used in the context of Web search to adapt search results for the needs of a *community* of like-minded users. In each case-study we will motivate the need for personalization, describe the personalization techniques that have been deployed, and outline some evaluation results that highlight the potential value of these techniques for the domain at hand.

It is worth noting that the research behind both case-studies has been published in separate articles previous to this. However in this article our focus is on contrasting these approaches to personalized information access and consider their pros and cons with respect to the different ways that they balance the personalization opportunity with users' likely privacy concerns.

1.2. Case-Study 1 — Personalized Mobile Portals

Mobile phones are now poised to overtake more traditional information access devices (e.g. desktop and laptop PCs) as the dominant platform for Internet information access in many markets. A recent report published by Strategy Analytics shows that mobile phone subscriptions approached 2.2 billion at the end of 2005 and are forecast to reach 2.5 billion by the end of 2006.[13] Indeed, at the time of writing, a new report by Ipsos Insight highlights how mobile phone ownership is now reaching saturation levels in many areas of the world. According to market research firm *Ipso Insight*,[11] globally, 28% of mobile phone subscribers have used their phones to browse the Internet (just over a 3% increase on 2004 figures). In Japan 40% of subscribers use the Mobile Internet while in Europe, France and the UK are leading the way; according to a 2005 Forrester Research study, 21% of European mobile subscribers use Mobile Internet services at least once per month, for example, refer to Ref. 35. Given this growth in Mobile Internet usage, and mobile information access in general, it is likely that the Mobile Internet will rapidly become a vital source of anytime, anywhere access to information for hundreds of millions of users.

However, there are problems. Mobile devices are not ideal information access devices, as we shall see. They have limited display capabilities and even more limited input features, all of which conspire to make mobile information access extra challenging from a usability viewpoint. In what follows we will describe some of these challenges in more detail, and outline their implications when it comes to mobile browsing. Particular attention will be paid to the so-called *click-distance* problem, which highlights how onerous it can be for users to navigate to content via mobile portals . We will go on to describe and evaluate how portals can be adapted in line with the browsing habits of users so as to significantly reduce the click-distance problem and lead to a corresponding increase in usability and usage.

1.2.1. *The challenges of mobile information access*

The Mobile Internet refers to the delivery of a variety of data services across wireless networks for Internet-enabled handsets, including Web-style information content, email services, games, maps, etc. Access devices range from limited, first-generation WAP (Wireless Application Protocol) phones to today's sophisticated PDAs and so-called smart phones. In the past, the usability of these mobile services has been compromised by limited device functionality, bandwidth and content. Today's new generation of 3G mobile services, however, represents a significant step forward so that major bandwidth and content issues have largely been resolved, with the latest phones offering users significant interface and functionality improvements over the original models. Nevertheless significant and challenging usability problems still remain. For example, mobile portals suffer from the prevailing *once-size-fits-all* approach to portal design that has been inherited from the traditional Web-space. The core problem being that the menu-driven nature of mobile portals (see Fig. 1.1), means that users spend a significant amount of their time online navigating to content through a series of hierarchical menus (their *navigation time*), and relatively limited time interacting with content (their *content time*). This limits the overall user experience since navigation time is essentially *valueless*, at least from a user's perspective.[23,27]

1.2.1.1. *Mobile internet devices*

From a user-experience viewpoint, one of the key features of the Mobile Internet is the degree to which existing consumer devices represent a significant step backwards in terms of their functionality, at least when compared to the traditional Internet device (the desktop PC or laptop). In particular, presentation and input capabilities tend to be extremely limited on most mobile devices. For instance, a typical desktop PC, with a screen size of 1024×768 pixels, offers more than 10 times the screen real-estate of a PDA, and more than 20 times the screen space of Internet phones (e.g. I-mode and Vodafone Live! handsets or Microsoft's SmartPhone).

Mobile handsets are further limited by their capacity to receive user input. The keyboard and mouse functionality of a modern PC are notably absent and the mobile phone numeric keypad makes it extremely difficult for user to input any quantity of information. From a Mobile Internet viewpoint, these devices restrict input features to simple scroll and select keys that allow the user to scroll through menu lists and perform simple selections. Some improvements are present in most PDAs, which tend to offer touch sensitive screens that are easier to manipulate. Nevertheless data input remains difficult at best.

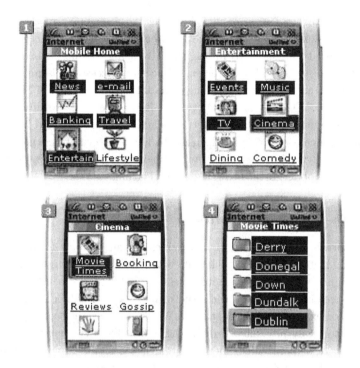

Fig. 1.1. Modern mobile portals offer users menu-driven access to content. The example screenshots show a sequence of menus from a portal home page to a local cinema site. In this scenario, the user is expected to navigate to their favourite cinema by scrolling to, and selecting, the appropriate intermediate menus (*Entertainment, Cinema, Movie Times, Dublin Cinemas*, etc.). A total of 15 clicks are ultimately required, made up of 4 menu selections and 11 menu scrolls.

1.2.1.2. Browsing versus search on the mobile internet

These differences that exist between mobile handsets and more traditional Internet devices, such as PCs and laptops, directly influence the manner in which users access information using these devices. For example, on the Internet today search has largely become the primary mode of information access. It is relatively easy for users to input search queries and search engines have improved significantly in their ability to respond intelligently to user needs. In addition the large screen sizes make it feasible for users to efficiently parse the longs lists of search results returned. In contrast, search is far more problematic on mobile devices. Entering queries is simply too time consuming and complex for the average user to tolerate and small screen sizes make it practically impossible for users to easily process the result lists returned. As a result, browsing is the primary mode of information access on the Mobile Internet. Instead of searching for information, users attempt to navigate to information by using mobile portals. Today the vast majority of Mobile Internet services are accessed via an operator portal with direct search constituting a small fraction (<10%) of Mobile Internet activity. This distinction between alternative modes of information access on the mobile and fixed Internet is an important one and it sets the scene for this case-study. To help users to locate information and services more effectively on the Mobile Internet we must attempt to improve the efficiency of mobile portal browsing or navigation.

1.2.2. The click-distance problem

Mobile portals are examples of hierarchical menu systems (HMS), which have a long history of research when it comes to understanding their general usability and navigation characteristics.[12,14,15,17,19,34,36–38] Much of this early research has focused on the structural properties of hierarchical menu systems, for example, their depth and width, as they relate to the ability of a user to easily navigate through the HMS. The evidence is clear: the complexity of a hierarchical menu system has a significant impact on its usability and the ability of users to navigate through menu levels. The type of menu hierarchies found on the Mobile Internet have been found to be subject to similar findings with portal complexity modeled in terms of the number of device interactions (i.e. menu selections and scrolls) needed for a user to locate a particular item of content.

The scale of the navigation problem associated with mobile portals today, and the mismatch between user expectations and realities, is highlighted by a number of early studies on mobile usability. For instance, Ref. 23 highlights how the average user expects to be able to access content within 30 seconds, while the reality was closer to 150 seconds at the time of the study. The time it takes a user to access a content item is a useful measure of navigation effort and the navigation effort associated with an item of content depends on the location of that item within the portal structure, and specifically on the number of navigation steps that are required in order to locate and access this item from the portal home page.

With most mobile phones today, there are two basic types of navigation action. The first is the *select*: the user clicks to select a specific menu option. The second is a *scroll*: the user clicks to scroll up or down through a series of options. Accordingly, an item of content, i, within a mobile portal can be uniquely positioned by the sequence of selects and scrolls needed to access it, and the navigation effort associated with this item can be simply modeled as click-distance, the corresponding number of these selects and scrolls [see Eq. (1.1)] . This simple model of click-distance has been shown to be a strong predictor of the navigation time associated with its access.[27,32,33]

$$\text{Click} - \text{Distance}(i) = \text{Selects}(i) + \text{Scrolls}(i) \qquad (1.1)$$

Recent studies illustrate the extent of the click-distance problem. For example, a recent analysis of 20 European mobile portals reports an average click-distance in excess of 16.[27] In other words, a typical European mobile user can expect to have to make 16 or more clicks (scrolls and selects) to navigate from their portal home page to a typical content target. Moreover, on average European portals are organized such that less than 30% of content sites are within 10–12 clicks of the portal home page; 10–12 clicks corresponds to a navigation time of about 30 seconds, which is expected by Mobile Internet users.[23] In other words the majority of mobile portal content was found to be essentially invisible to users because of its positioning within its parent portal.

1.2.3. *Personalized navigation*

One way to relieve the navigation problem is to reduce the click-distance of a portal (see Ref. 20). But large click-distances are a fundamental feature of a *one-size-fits- all* approach to portal design and optimizing a menu structure for the needs of some imaginary "average user" is unlikely to benefit individual users. However, a solution is at hand: instead of presenting the same portal to each and every user it is possible to use user profiling and personalization techniques to learn about the preferences of individual users in order to strategically adapt the structure of the portal on a user by user basis.[2,3,8,21,22,24,31–33] Each time a user accesses a given menu, m, this menu is dynamically created based on their short and long-term behavior. Importantly, the click-distance of the content items that a user is likely to be interested in is reduced by promoting these items to higher menus within the portal structure.

1.2.3.1. *Profiling the user*

Tracking user accesses across a mobile portal provides the basis for an effective profiling mechanism. Individual menu accesses are stored in a so-called *hit-table*, which provides a snapshot of a user's navigation activity over time. For example, Fig. 1.2 indicates that a user has accessed option B from menu A 10 times and option C 90 times; of course in reality other activity information including device, temporal and location information is normally stored as part of this evolving profile, but a more detailed discussion is outside of the scope of this paper.

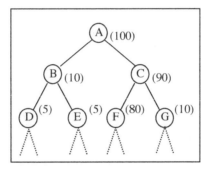

Fig. 1.2. The user hit-table reflects portal access frequencies at the level of an individual user. One such hit-table is stored for each unique user as a record of their access behaviour.

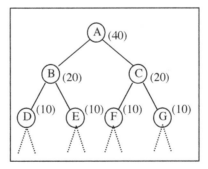

Fig. 1.3. The static hit-table is used to encode a default portal structure. A single-static hit-table is maintained across all portal users and the relative values of the hit-table nodes provide one way to control the sensitivity of the portal to the activities of users.

In fact two types of hit-table are used: a global, *static* hit-table that is initialized with respect to the default portal structure [Fig. 1.3]; and a *user* hit-table that records each user's individual history. The static table makes it possible to deliver a default menu structure during the early sessions. This will be over-ridden by the personalized menu, during later sessions, once a user's access probabilities build-up as a result of their activity. Moreover, the hit values set in the static table make it possible to control personalization latency — low values mean that personalization takes effect very quickly.

1.2.3.2. *Personalizing the portal*

To build a personalized menu m we must identify the k most probable options for m (the k options with the highest $P_u(o|m)$ values — the probability that user u will access option o given that they are in menu m) using the frequency information in the user and static hit-tables. Consider the data in Fig. 1.2 and Fig. 1.3] wit respect to the construction of menu A. The access probabilities can be determined as shown in Fig. 1.4. In descending order of access probability we have C, F, B, G, D, and E. For $k = 3, C, F$ and B are selected, in order, for menu A.

The complexity of the proposed personalization method depends on the complexity of the process that identifies the k most probable options for the menu, m. As described this can mean examining not just the default options of m, but also all the options contained in menus that are descendants of m; essentially a breadth-first search from m to the content leaves of the menu tree is required. Fortunately, a more efficient algorithm is possible once we recognize that, by definition, $P_u(o|m)$ is always greater than or equal to $P_u(o'|m)$ where o' is an option of a menu, m', which is itself a descendent of m through o. This means that we can find the k most probable nodes for menu m by performing a depth-limited, breadth-first search over the menu tree rooted at m. We only need to expand the search through an option o' if $P_u(o'|m)$ is greater than the kth best probability so far found. Once again, a detailed description of this issue is beyond the scope of the current

$P(B|A)$	$= (20+10)/(40+100)$	$= 0.214$
$P(C|A)$	$= (20+90)/(40+100)$	$= 0.786$
$P(D|A) = P(B|A)P(D|B)$	$= (30/140)(10+5)/(20+10)$	$= 0.107$
$P(E|A) = P(B|A)P(E|B)$	$= (30/140)(10+5)/(20+10)$	$= 0.107$
$P(F|A) = P(C|A)P(F|C)$	$= (110/140)(10+80/20+90)$	$= 0.642$
$P(G|A) = P(C|A)P(G|C)$	$= (110/140)(10+10)/(20+90)$	$= 0.142$

Fig. 1.4. Sample access probabilities.

paper but the interested reader is referred to Refs. 32 and 33 for further information.

The approach just described supports two types of menu adaptations called *vertical promotions*. A menu option may be promoted *within* its parent menu; that is, its relative position within the parent menu is adjusted. A promotion *between* menus occurs when an option is promoted into an ancestral menu. Of course promotions are side-effects of the probability calculations. In the above example, option F is promoted to A's menu — options can even be promoted from deeper levels if appropriate — because it is a probable destination for this user. If F is subsequently selected from A, it is added to A's hit table entry for that user, so the next time that A is created, the computation of $P(F|A)$ must account for the new data on F. Specifically, assuming a single access to F as an option in A, we get:

$$P(F|A) = 1/101 + (110/141)(10 + 80/20 + 90) = 0.647.$$

An example of a portal after personalization has occurred is presented in Fig. 1.5. In this case, the original portal presented in Fig. 1.1 has been adapted, based on their usage history, to provide more direct access to local cinema listings. In this example, the *Entertain* option on the home page has been promoted to be the top option and their local cinema's site (*Ster Century*) has been promoted out of the *Movie Times* section of the portal and into the *Entertain menu*.

1.2.4. *Evaluation*

We believe that user satisfaction will be significantly enhanced by using personalization to reduce the click-distance of a mobile portal, on a user by user basis. In turn, by enhancing user satisfaction we expect to experience a significant increase in portal usage as users come to recognize the value of the Mobile Internet. The evaluation presented in this section is part of a set of live-user field trials carried out on European mobile portals. In this case the trial consisted of a two-week profiling period in which no personalization took place, but the behavior of the users was monitored in order to profile their navigation patterns. The remaining four weeks were divided into two two-week personalization periods. During this time profiling continued but in addition, personalization was switched on so that users experienced a new portal structure that was adapted to their navigation preferences. The reported trial consists of 130 trialists from a variety of backgrounds and with a range of mobile usage habits and handsets.

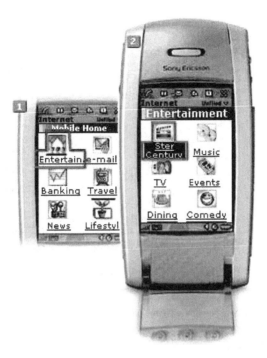

Fig. 1.5. Screenshots showing a portal after it has been personalized to provide the user with more direct access to their local cinema listings. Note that the *Entertainment* menu in the portal home page has now been promoted to the top of the page and that the *Ster Century* local cinema option has been promoted from the depths of the entertainment section of the portal right to the top of the main entertainment menu. Due to the consistent past behavior of this example user, their local cinema is now within 2 clicks of the portal home page, instead of the 15 clicks required to access this service in its default portal position.

1.2.4.1. *Click-distance reduction*

Figure 1.6 illustrates how portal click-distance is altered during the trial as a result of personalization. In this instance click-distance is measured in terms of the average click-distance from the home page to each user's top three most frequently accessed sites. In summary, the results show how a starting click-distance of 13.88 for the static portal drops by over 50% to 6.84 during the first personalization period and by a further 2% for the final period. These results show two things: first that significant click-distance reductions are possible; and second, these reductions are realized very rapidly, in this case after only two weeks of profiling, which corresponds to about 3–5 sessions per user, there is a marked impact on click-distance and thus, we assume, navigation effort.

Click-Distance

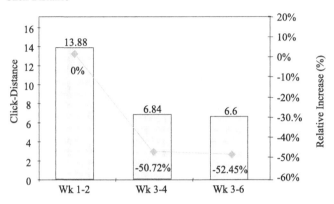

Fig. 1.6. Click-distance results.

Average per-Site Navigation Time per User

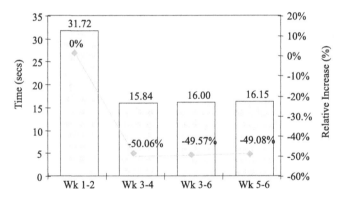

Fig. 1.7. Navigation time per site access.

1.2.4.2. *Navigation time versus content time*

Navigation effort is considered explicitly in Fig. 1.7, which shows the time taken by a typical user navigating to a typical content site; this was obtained by computing the average navigation time per session and dividing by the average number of sites visited per session. Navigation time is seen to decrease significantly during the trial, as the portal adjusts to each user's

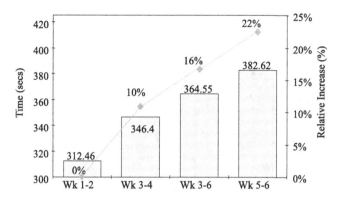

Fig. 1.8. Content time results.

habits. During the control period, the average trialist was taking nearly 32 seconds to navigate to a typical content site. However, this falls to about 16 seconds as a result of personalization, a relative reduction of about 50%. In contrast, when we look at the daily content time for users (Fig. 1.8) we find that there is a significant increase due to personalization. Over the four-week personalization period (weeks 3–6) average daily content time increases by over 16% overall. During the static period the average total daily content time per trialist is about 312 seconds compared to over 364 seconds as an average of the four-week personalization period. Moreover, if we look at the average content time for the final two trial weeks (as opposed to the final four weeks) we find a relative increase of more than 22% (average content time of nearly 383 seconds). Thus, the relative increase in content time for the final two weeks of the trial (22.45%) has more than doubled in comparison to the first two weeks of personalization (10.89%); as personalization, proceeds so too do the usage benefits increase.

Of course mobile operators have largely moved from time-based Mobile Internet charging models towards content-based models. The results in Fig. 1.9 show that this extra content time arises because users are downloading additional content by visiting extra sites and services. For example, we see that users are visiting between 24% and 28% extra content sites per day as a result of personalization, which has significant revenue implications for mobile operators.

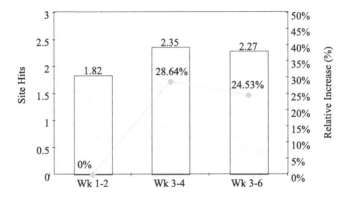

Fig. 1.9. Site-hit results.

1.3. Case-Study 2: Personalized Web Search

For our second case-study we look at information access on the Web, focusing on Web search as the primary way for people to locate Web-based information services today. We are all familiar with query-based Web search and we have all participated in the rapid growth of search engines over the past 10 years. Recent statistics by search engine marketing service, *Search Engine Watch (www.searchenginewatch.com)*, estimates that in the United States alone, there were in the region of 200 million searches per day during March 2006. However, although today's search engines have evolved from their information retrieval roots, they continue to struggle when it comes to delivering the right information at the right time to the right user. In this case-study we will explore some of the challenges facing Web search engines and explain how one particular approach to personalization can have a significant benefit when it comes to delivering relevant results to communities of like-minded searchers.

1.3.1. *The challenges of web search*

Web search is challenging for a variety of reasons. For a start the sheer scale and heterogeneity of the Web represents a significant information access challenge in and of itself. Recent estimates of the Web's current size speak about a rapidly growing, distributed and diverse repository of 10 seconds of billions of publicly accessible information items, from the largely text-based content of HTML Web pages, PDFs and blogs to less structured content such as photos, video and podcasts. Recent studies suggest a growth rate that tops 60 terabytes of new information per day:[25] in 2000 the entire World-Wide Web consisted of just 21 terabytes of information, by 2004, it was growing by three times this figure every single day.[18]

Web search is made all the more difficult because of the nature of Web searchers and their queries. Today's typical Web searcher is a far cry from the information retrieval (IR) expert contemplated by the IR engines that lie at the core of modern search engines. Web searchers rarely produce high quality queries: typically they are vague and ambiguous, with the average query containing only about two query terms.[16] For example, consider a query like *"jordan pictures"* which offers no clues about whether the searcher is likely to be looking for images of the Formula One racing team, the middle eastern state, the basketball star, or the UK model (a.k.a. Katie Price). Moreover, people use a wide variety of terms to refer to the same types of information[9] and as a result there is often a mismatch between the terms found in search queries and the terms found within the documents being sought: this lack of a strong correspondence between the query-space and the document-space introduces a major problem for the term-based matching approaches that form the kernel of modern search engines.[4]

Our research has focused on addressing the *vague query problem* and the *vocabulary gap* that plagues modern Web search. Our approach has been to look at how a variety of artificial intelligence inspired techniques can be used to recast the traditional document-centric view of Web search as one that emphasizes the vital role that Web searchers themselves can play in solving the search problem. We combine ideas from case-based reasoning, user modeling, and information filtering with a view to developing Web search engines that leverage the *social power* of the Web. In short, we argue that it is useful to think of Web search as a social activity in which *ad hoc* communities of like-minded searchers tend to search for similar types of information in similar ways. And we demonstrate that by capturing the search experience of such communities it is possible to adapt traditional (general-purpose) search engines so that they can respond more effectively to the needs of different communities of searchers, even in the face of vague queries. For example, when a member of a motoring community is searching for *"jordan pictures"* she is likely to select results related to the Formula One racing team, and the past search behavior of other community members should support this.

In this case-study we outline our work on a community-based approach to Web search known as *Collaborative Web Search (CWS)* ; see Refs. 28

and 30. CWS is a post-processing (meta-search) technique that maintains a profile of the search patterns and preferences of separate communities of searchers. When responding to a new query by some community member, CWS uses the host community's profile to enrich the results returned by an underlying search engine(s) by identifying and promoting results that have been previously selected by community members in response to similar queries. After motivating and describing the core CWS technique we go on to summarize recent results that highlight the potential for CWS to significantly improve the precision of the results returned by the underlying search engine.

1.3.2. *Exploiting repetition and regularity in community-based web search*

Collaborative Web search is motivated by regularity and repetition that is inherent in Web search, especially among the searches of communities of like-minded individuals: similar queries tend to recur and similar pages tend to be selected for these queries. CWS proposes to exploit these regularities when responding to new queries by reusing the result selections from similar past queries.

How commonplace is community-based search and how regular are community search patterns? Even though most searches are conducted through generic search engines many are examples of community-based searches. For instance, the use of a Google search box on a specialized Web site (e.g. a motoring enthusiast's site) suggests that its searches are likely to be initiated by users with some common (motoring) interest. Alternatively, searches originating from a computer laboratory assigned to second year students are likely to share certain characteristics related to their studies (courses, projects etc.) and social lives (societies, gigs etc.).

Previous analyses of search engine logs have shown how query repetition and selection regularity is prevalent in community oriented search scenarios. For example, Ref. 30 reports how up to 70% of search queries from community searches share at least 50% of their query terms with other queries. Moreover, they show that there is a strong regularity between the selections of community members in response to similar queries: similar queries lead to similar selections. CWS takes advantage of this repetition and regularity by recording community searches (queries and result selections) and then promoting results that have been regularly selected in the past by community members in response to similar queries to the target.

1.3.3. *A case-based approach to personalizing web search*

The basic CWS architecture is presented in Fig. 1.10. Briefly, when a new target query q_T is submitted, in the context of some community, result-list R_T is produced from the combined results of the underlying search engines $(S_1, \ldots, S_n)$, R_M plus a set of promoted results (R_P) chosen because they have been previously selected by community members for queries that are similar to the target.

Collaborative Web search adopts a case-based reasoning perspective[1,5] in the sense that past search experiences are harnessed to help respond to new target search queries. The search history of a given community is stored as a case-base of search cases with each search case made up of a specification part and a solution part; see Eq. (1.2). The *specification* part [see Eq. (1.3)] corresponds to a given query. The *solution* part [see Eq. (1.4)] corresponds to a set of selection-pairs; that is, the set of page selections that have been accumulated as a result of past uses of the corresponding

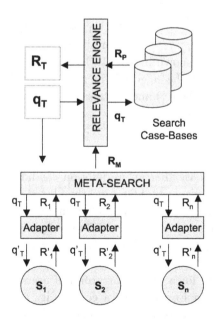

Fig. 1.10.   The collaborative web search architecture.

query. Each selection-pair is made up of a result-page id and a hit-count representing the number of times that the given page has been selected by community members in response to the given query.

$$c_i = (q_i, (p_1, r_1), \ldots, (p_k, r_k)) \tag{1.2}$$

$$\text{Spec}(c_i) = q_i \tag{1.3}$$

$$\text{Sol}(c_i) = ((p_1, r_1), \ldots, (p_k, r_k)). \tag{1.4}$$

Given a new target query, q_T, CWS must identify a set of *similar* search cases from the community's search case-base. A standard term-overlap metric [Eq. (1.5)] is used to measure query-case similarity, to rank-order past search cases according to their target similarity, so that all, or a subset of, similar cases might be reused during result ranking.

$$\text{Sim}(q_T, c_i) = \frac{|q_T \cap \text{Spec}(c_i)|}{|q_T \cup \text{Spec}(c_i)|}. \tag{1.5}$$

Consider a page, p_j, that is associated with query, q_i in some search case, c_i. The relevance of p_j to c_i can be estimated by the relative number of times that p_j has been selected for q_i; see Eq. (1.6).

$$\text{Rel}(p_j, c_i) = \frac{r_j}{\sum_{\forall r_m \in \text{Sol}(c_i)} r_m}. \tag{1.6}$$

Then, the relevance of p_j to some new target query q_T can be estimated as the combination of $\text{Rel}(p_j, c_i)$ values for all cases $c_1, \ldots, c_n$ that are deemed to be similar to q_T, as shown in Eq. (1.7). Each $\text{Rel}(p_j, c_i)$ is weighted by $\text{Sim}(q_T, c_i)$ to discount the relevance of results from less similar queries; $\text{Exists}(p_j, c_i) = 1$ if $p_j \in \text{Sol}(c_i)$ and 0 otherwise.

$$W\text{Rel}(p_j, q_T, c_1, \ldots, c_n) = \frac{\sum_{i=1 \cdots n} \text{Rel}(p_j, c_i) \bullet \text{Sim}(q_T, c_i)}{\sum_{i=1 \cdots n} \text{Exists}(p_j, q_i) \bullet \text{Sim}(q_T, c_i)}. \tag{1.7}$$

This weighted relevance metric is used to rank-order search results from the community case-base that are promotion candidates for the new target query. The top ranked candidates are then listed ahead of the standard meta-search results to give R_T; see Ref. 6 for further details on the search interface and result presentation. The CWS technique has been implemented in the I-SPY (http://ispy.ucd.ie) search engine, which allows groups to create and harness search communities in order to benefit from search results that are personalized for a particular group of users. An example search session is presented in Fig. 1.11 for a community that was created by a local Dublin software company specializing in the development of personalization technology. The screen-shot shows the top results for the query

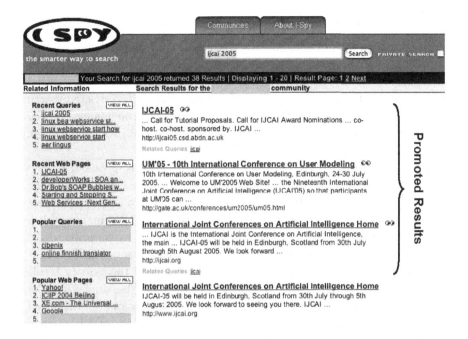

Fig. 1.11. An example screenshot showing promoted results.

"*IJCAI 2005*", referring to the main artificial intelligence (AI) conference, and the top three presented results have been promoted by the CWS engine. These results have been promoted because they have been selected for similar queries, by other searchers in this community, in the past. It is worth commenting on the second promoted result. This result refers to a related conference on user modeling and normally this result would not have been so prominent in the ranking for this query. However, many of the community members, given their interest in personalization, also have an interest in user modeling and in 2005, both conferences took place in Edinburgh back to back. Hence, this promotion makes particular sense to this community of searchers and is a good example of the type of relevant result that might ordinarily be missed.

1.3.4. *Evaluation*

Previous evaluations of collaborative Web search have included a mixture of artificial-user and live-user studies.[7,29,30] However these studies have been limited in various ways, either because they have involved artificial user models or because they have used live users but in a restricted search scenario. In this section we review the results of a more open-ended search trial involving 50 staff from a local software company over a four-week period in late 2005.

During the trial the participants were asked to use I-SPY as their primary search engine; prior to the trial 90% of search sessions used Google. The I-SPY system was setup to use Google and HotBot as its underlying search engines and a new community was created for participants with a hit-matrix initialized from search log data for the nine weeks prior to the start of the trial. I-SPY's query-similarity threshold was set at 50%, so that only those past sessions that shared more than half of their query terms with the current target query would be considered to be similar for the purposes of result promotion (see Sec. 1.3.3). Participants were introduced to I-SPY via a short explanatory email and encouraged to use it as they would a normal search engine. Over the four weeks more than 1500 queries were submitted and more than 1800 result URLs were selected.

1.3.4.1. *Successful sessions*

The real test of the system is whether I-SPY's promotions turn out to be relevant for the searcher than the default Google and HotBot results. Evaluating the relevance of search results in a trial such as this is difficult to do in a direct fashion. For example, standard search interfaces do not provide a facility to allow users to indicate how well their information needs have been met by search results, and while it would be possible to add such a facility to I-SPY for the purpose of measuring relevance in this trial, many users indicated that they would find this to be a nuisance. For this reason we examine a less direct measure of relevance. Instead we propose that the selection of at least one result in a given search session acts as a crude, but nevertheless useful indicator of result-list relevance. We refer to a search session, where at least one result has been selected, as a *successful session*. If no results are selected (a *failed session*) then we can be relatively confident that the search engine has not retrieved a result that is *obviously* relevant to the searcher. Note that we do not distinguish here between sessions with different numbers of selected results, mainly because it is not possible to

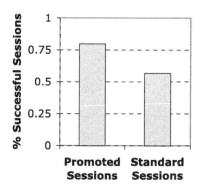

Fig. 1.12. A comparison of the success rates of standard and promoted search sessions.

conclude much from the frequency of result selections. For example, one might be tempted to conclude that users selecting more results is a sign of increasing result relevance, except that a similar argument can be made in support of decreasing result relevance, on the basis that the initial selections must not have satisfied the users.

To analyze the ability of collaborative search to deliver successful sessions, we split the search sessions into those that contained promotions (*promoted sessions*) and those that did not (*standard sessions*). The former correspond to sessions where collaborative search has the potential to influence relevance, whereas the latter serve as a pure meta-search benchmark against which to judge this influence. Incidentally, there appears to be no difference between the queries for the promoted sessions when compared to those for standard sessions and both sets of queries have almost identical distributions; for example, an average of 2.4 terms per query for the promoted sessions compared to 2.5 for the standard sessions was measured. Indeed, given enough time it is likely that many of the standard queries would eventually be paired with new similar queries and so participate in future promoted sessions.

Figure 1.12 presents the average percentage of successful sessions among the promoted and standard sessions and demonstrates a clear advantage for the promoted sessions. On average, 80% of the promoted sessions were successful, compared to 56% for the standard sessions, a difference that is significant at the 99% confidence level. In other words, the collaborative search result-promotion mechanism leads to a 40% relative improvement in the chances that a given search will translate into a successful search session.

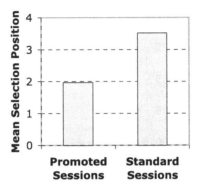

Fig. 1.13. A comparison of the mean selected position of selected results in standard and promoted search sessions.

1.3.4.2. *Selection positions*

As a complementary measure of result-relevance, it is also interesting to compare the promoted and standard sessions in terms of the average position of selected results within successful sessions; that is, those sessions in which selections have been made. We would like to see relevant results appearing higher up in result-lists. Moreover, assuming that users are likely to select results that at least appear to be more relevant than those that do not, then we would like to minimize the mean position of a selected result.

Figure 1.13 presents the mean position of the selected results among the successful sessions of the promoted and standard sessions. This once again shows a clear advantage for the former. On average, the mean position of a selected result among the successful promoted sessions is 1.96, compared to 3.51 for the successful standard sessions. This difference is statistically significant at the 99% confidence level and corresponds to a 44% reduction in the position of relevant results for promoted sessions compared to standard sessions.

It is worth commenting on the importance of this observed difference in the selection positions. While there is an advantage due to the promoted sessions, one might ask whether the observed reduction of one or two places is likely to be important. We believe that it is, for a number of reasons, not the least of which is that results should be ordered by their expected relevance as a matter of course. In addition, users have a tendency to focus their attention on the top-ranked results. The fact that promoted sessions have a higher success rate than the standard sessions is likely due to this difference in the position of apparently relevant results, because for the most part I-SPY promotes results from lower-down in the standard result-lists (returned by Google and HotBot) to higher positions. Moreover, the observed difference may become even more important in other search scenarios, such as mobile search, where screen-space is so restricted as to severely limit the number of results that may be presented on a single screen.

1.4. Personalization-Privacy: Striking a Balance

Personalization technologies are undoubtedly an important component in any practical solution to the information overload problem. By understanding our preferences and likely interests, personalized information services can do a far better job when it comes to predicting and responding to our current needs. The upside of personalization is a more efficient information access experience for end users. For example, in this paper we have explored two case-studies, from different domains (the Mobile Internet versus the traditional Web), and involving different modes of information access (browsing versus search). In each scenario, the availability of personalization technology was seen to improve the end-user experience, either by helping users to navigate more efficiently to content on the Mobile Internet, or by ensuring that search results that are likely to match a user's unstated preferences are promoted, in the case of Web search.

There is, of course, a downside to personalization: users must be willing to sacrifice some level of privacy if their preferences are to be learned, especially when these profiles are stored on the server-side. In our experience, users view personalization as a privacy tradeoff and the degree to which they are willing to trade their personal information for a degree of personalization depends greatly on the type of information that is requested and the nature of the improvement that they expect to enjoy as a result. Our two case-studies serve to highlight two very different approaches to this personalization-privacy tradeoff.

In the first case-study the personalization techniques employed were designed to capture the preferences of the individual. However, the nature of the profiled information is less likely to be of concern to the individual user, after all we were only storing information about their navigation patterns through a set of fixed portal categories. This seems to be borne out in practice because in deployment scenarios the vast majority of users have indicated a willingness to opt-in to portal personalization. In this sense it appears that most users are willing to accept a limited set of personal preferences (related to preferred portal content categories) for a greatly improved mobile portal service.

The situation appears to be somewhat different in the area of Web search however; for example, at the time of writing (June 2007) serious questions were being asked of Google in relation to its privacy policies and the practice of recording and storing the long-term search profiles of individual users. Certainly storing the individual queries submitted by searchers, and the results each searcher selects, is likely to raise privacy concerns in the mind of the searcher. Our search patterns are far more revealing that our browsing patterns, especially when browsing is limited to a well-defined portal. The queries we submit to search engines speak volumes about our likes and dislikes on as wide a range of topics as one could imagine. This, after all, is why Google and Yahoo have built such a successful advertising business out of search. For these reasons, when looking to apply personalization techniques to Web search, we felt it inappropriate to consider profiling the individual. Instead, our Web search case-study describes an anonymous form of profiling, where the search behavior of an individual is not recorded in a way that can be traced back to the individual. Instead, it is merged with the relevant community's search experience, disguising the input of any individual, but still facilitating a level of personalization that can help to improve the user's search experience.

In summary, we argue that in both case-studies a reasonable balance has been struck with respect to personalization and privacy. The individual user profiles in the first case-study represent a minor sacrifice in a user's privacy set against the significant navigation benefits that were found as a result. In recent years the technology described in this case-study has been widely rolled out across European mobile operators by ChangingWorlds (www.changingworlds.com) and has met with extremely high adoption rates (approximately 85%) among Mobile Internet subscribers, suggesting that users recognize their navigation patterns being profiled as fair trade considering the navigation benefits they enjoy as a result. On the other hand, the community-based profiles used in the second case-study provide for a level of profiling that is virtually anonymous from a user's perspective, whilst still providing for a level of personalization that can deliver value to the individual searcher. As the search industry looks to personalization technologies to drive the next generation of search engines, we believe that users will come to resist server-side profiling if it means that detailed individual profiles are being maintained. Instead we argue that less invasive profiling, such as community-based profiling, provides a more practical compromise.

1.5. Conclusions

In this paper we have highlighted the role to be played by personalization technologies when it comes to relieving the information overload problem. Today, personalization technologies are becoming more commonplace as information providers and users alike come to appreciate the benefits on offer. In this paper, we have reviewed two case-studies of the use of personalization in mobile browsing and Web search as a basis for describing two very different approaches to personalization. However, the potential benefits of personalization come at a price and there is always a cautionary note regarding the privacy of individuals when it comes to recording and reusing information about their online behavior or information preferences. In our case-studies we have highlighted how the different approaches to personalization have different implications when it comes to user privacy. In short, getting personalization right means striking a balance between what we *can* learn about users and what we *should* learn about users. If we can get this balance right the practical benefits of personalization will become a reality. If we get this balance wrong, and usually this means *over profiling* the user, then personalized information services will not succeed in the marketplace.

Acknowledgments

This material is based in part on works supported by ChangingWorlds Ltd., and on works supported by Enterprise Ireland's Informatics Initiative and Science Foundation Ireland under Grant No. 03/IN.3/I361.

References

1. A. Aamodt and E. Plaza, Case-based reasoning: foundational issues, methodological variations, and system approaches, *AI Commun.* **7**(1) (1994) 39–59.

2. C. Anderson, P. Domingos and D. Weld, Adaptive web navigation for wireless devices, *Proc. 17th Int. Joint Conf. Artificial Intelligence* (2001), pp. 879–884.
3. D. Billsus, M. J. Pazzani and J. Chen, A learning agent for wireless news access, *Proc. Conf. Intelligent User Interfaces* (2000), pp. 33–36.
4. P. Bollmann-Sdorra and V. V. Raghavan, On the delusiveness of adopting a common space for modeling ir objects: Are queries documents? *J. Amer. Soc. Inform. Sci.* **44**(10) (1993) 579–587.
5. R. L. de Mantaras, D. McSherry, D. Bridge, D. Leake, B. Smyth, S. Craw, B. Faltings, M. L. Maher, M. T. Cox, K. Forbus, M. Keane, A. Aamodt and I. Watson, Retrieval, reuse, revision and retention in case-based reasoning, *Knowl. Engin. Rev.* **20**(3) (2006) 215–240.
6. J. Freyne and B. Smyth, Cooperating search communities, in *Proc. 4th Int. Adaptive Hypermedia and Adaptive Web-Based Systems*, Dublin, Ireland (2006), pp. 101–110.
7. J. Freyne, B. Smyth, M. Coyle, E. Balfe and P. Briggs, Further experiments on collaborative ranking in community-based web search, *Artif. Intell. Rev.: An Int. Sci. Engin. J.* **21**(3–4) (2004) 229–252.
8. X. Fu, J. Budzik and K. Hammond, Mining navigation history for recommendation, *Proc. Conf. Intelligent User Interfaces* (2000), pp. 106–112.
9. G. W. Furnas, T. K. Landauer, L. M. Gomez and S. T. Dumais, The vocabulary problem in human-system communication, *Commun. ACM* **30**(11) (1987) 964–971.
10. D. Godoy and A. Amandi, PersonalSearcher: an intelligent agent for searching web pages, in *IBERAMIA-SBIA*, **1952** (Springer, 2000), pp. 62–72.
11. IposInsight, Mobile phones could soon rival the PC as world's dominant internet platform, http://www.ipsos-na.com/news/pressrelease.cfm?id=3049 (April 2006). Last checked June 8th, 2006.
12. J. Jacko and G. Salvendy, Hierarchical menu design: dreadth, depth and task complexity, *Percep. Motor Skills* **82** (1996) 1187–1201.
13. P. Kendall, Worldwide cellular user forecasts, 2005–2010 (2006).
14. J. I. Kiger, The depth/breadth tradeoff in the design of menu-driven interfaces, *Int. J. Man/Mach. Stud.* **20** (1984) 201–213.
15. K. Larson and M. Czerwinski, Web page design: implications of memory, structure and scent for information retrieval, in *Proc. CHI'98 Human Factors in Computer Systems* (ACM Press, 1998), pp. 25–32.
16. S. Lawrence and C. L. Giles, Context and page analysis for improved web search, *IEEE Internet Comput.* (1998) 38–46.
17. E. Lee and J. MacGregor, Minimizing user search time in menu retrieval systems, *Human Factors* **27** (1985) 157–162.
18. P. Lyman and H. R. Varian, How much information, Retrieved from http://www.sims.berkeley.edu/how-much-info-2003 on January 14th, 2005, 2003.

19. D. P. Miller, The depth/breadth tradeoff in hierarchical computer menus, *Proc. 25th Ann. Meeting of the Human Factors and Ergonomics Society* (1981), pp. 296–300
20. M. Miyamoto, T. Makino and T. Uchiyama, Menu design for cellular phones, *Proc. Workshop on Mobile Personal Information Retrieval (SIGIR'02)* (2002).
21. M. Perkowitz, *Adaptive Web Sites: Cluster Mining and Conceptual Clustering for Index Page Synthesis*, Ph.D. thesis, Department of Computer Science and Engineering, University of Washington (2001).
22. M. Perkowitz and O. Etzioni, Towards adaptive web sites: conceptual framework and case study, *J. Artif. Intell.* **18**(1–2) (2000) 245–275.
23. M. Ramsey and J. Nielsen, *The WAP Usability Report*, Neilsen Norman Group (2000).
24. D. Reiken, Special issue on personalization, *Commun. ACM* **43**(8) (2000).
25. W. Roush, Search Beyond Google, *MIT Technol. Rev.*, 2004, pp. 34–45.
26. U. Shardanand and P. Maes, Social information filtering: algorithms for automating "word of mouth", in *Proc. Conf. Human Factors in Computing Systems (CHI '95)* (ACM Press, 1995, NY), pp. 210–217.
27. B. Smyth, *The Plight of the Mobile Navigator*, MobileMetrix (2002).
28. B. Smyth, E. Balfe, O. Boydell, K. Bradley, P. Briggs, M. Coyle and J. Freyne, A live-user evaluation of collaborative web search, in *Proc. 19th Int. Joint Conf. Artificial Intelligence (IJCAI '05)*, Edinburgh, Scotland (2005), pp. 1419–1424.
29. B. Smyth, E. Balfe, P. Briggs, M. Coyle and J. Freyne, Collaborative web search, in *Proc. 18th Int. Joint Conf. Artificial Intelligence, IJCAI-03* (Morgan Kaufmann, 2003, Acapulco, Mexico), pp. 1417–1419.
30. B. Smyth, E. Balfe, J. Freyne, P. Briggs, M. Coyle and O. Boydell, Exploiting query repetition and regularity in an adaptive community-based web search engine, *User Modeling and User-Adapted Interaction* **14**(5) (2004) 383–423.
31. B. Smyth and C. Cotter, Wapping the web: a case-study in content personalization for WAP-enabled devices, *Proc. 1st Int. Conf. Adaptive Hypermedia and Adaptive Web-Based Systems (AH'00)* (2000), pp. 98–108.
32. B. Smyth and P. Cotter, Personalized adaptive navigation for mobile portals, in *Proc. 15th European Conf. Artificial Intelligence — Prestigious Applications of Artificial Intelligence* (IOS Press, 2002), pp. 608–612.
33. B. Smyth and P. Cotter, The plight of the navigator: solving the navigation problem for wireless portals, in *Proc. 2nd Int. Conf. Adaptive Hypermedia and Adaptive Web-Based Systems (AH'02)* (Springer-Verlag, 2002), pp. 328–337.
34. K. Snowberry, S. R. Parkinson and N. Sisson, Computer display menus, *Ergonomics* **26** (1983) 699–712.
35. N. van Veen, M. de Lussanet and L. Menke, European Mobile Forecast: 2005 to 2010 (2005).

36. D. Wallace, N. Anderson and B. Shneiderman, Time stress effects on two menu selection systems, *Proc. 31st Ann. Meeting of the Human Factors and Ergonomics Society* (1987), pp. 727–731.
37. J. M. Webb and A. F. Kramer, Maps or analogies? A comparison of instructional aids for menu navigation, *Human Factors* **32** (1990) 251–266.
38. P. Zaphirs, Depth versus breadth in the arrangement of web links, *Proc. 44th Ann. Meeting of the Human Factors and Ergonomics Society* (2000), pp. 139–144.

BIOGRAPHY

Barry Smyth received a B.Sc. in computer science from University College Dublin in 1991 and a Ph.D. from Trinity College Dublin in 1996. He is currently the Head of the School of Computer Science and Informatics at University College Dublin where he holds the Digital Chair in Computer Science. He has published over 200 scientific articles in journals and conferences and has received a number of international awards for his research. His research interests include artificial intelligence, case-based reasoning, information retrieval, and user profiling and personalization.

Chapter 2

A Deep Evaluation of Two Cognitive User Models for Personalized Search

Fabio Gasparetti and Alessandro Micarelli

Department of Computer Science and Automation
Artificial Intelligence Laboratory
Roma Tre University
Via della Vasca Navale, 79 00146 Rome, Italy
gaspare@dia.uniroma3.it
micarel@dia.uniroma3.it

Personalized retrieval of documents is a research field that has been gaining interest, since it is a possible solution to the information overload problem. The ability to adapt the retrieval process to the current user needs increases the accuracy and reduces the time users spend to formulate and sift through result lists.

In this chapter we show two instances of user modeling. One is based on the human memory theory named Search of Associative Memory, and a further approach based on the Hyperspace Analogue to Language model. We prove how by implicit feedback techniques we are able to unobtrusively recognize user needs and monitor the user working context. This is important to provide personalization during traditional information retrieval and for recommender system development.

We discuss an evaluation comparing the two cognitive approaches, their similarities and drawbacks. An extended analysis reveals interesting evidence about the good performance of SAM-based user modeling, but it also proves how HAL-based models evaluated in the Web browsing context shows slightly higher degree of precision.

2.1. Introduction

Web users follow two predominant paradigms to access information: browsing and searching by query. In the first paradigm, users analyze Web pages one at a time, surfing through them sequentially, following hyperlinks. This is a useful approach to reading and exploring the contents of a hypertext, but it is not suitable for locating a specific piece of information. The larger the hypertextual environment is, the more difficulty a user will have finding what he is looking for.

The other information access paradigm involves querying a search engine, an effective approach that directly retrieves documents from an index of millions of documents in a fraction of a second. This approach is based on a classic Information Retrieval (IR) model: wherein documents and information needs are processed and converted into ad-hoc representations. These representations are then used as input to some similarity function that produces lists of documents.

In this paradigm, users are usually forced to sift through long result lists to find relevant information. Moreover, most of the times, the same result list is returned for the same query, regardless of who submitted the query, despite users usually having different needs. For these reasons, the identification of the user's information needs and the personalization of the human-computer interaction are becoming fundamental research topics.

The acquisition of user knowledge and preferences is one of the most important problems to be tackled in order to provide effective personalized assistance. Learning techniques for user modeling can be partitioned by the type of input used to build the profile. Explicit feedback systems rely on direct user intervention, that usually suggests keywords or documents of interest, or answers to questions about his/her needs.

Even though explicit feedback techniques have been shown to improve retrieval performance, some studies have found that these techniques are not able to considerably improve the user model,[1] especially if interfaces provided to manage the model are not very powerful.[2] Users are usually unwilling to spend extra effort to explicitly specify their needs, and are often not able to use those techniques effectively,[3,4] or might find them confusing and unpredictable.[5] Moreover, research shows that users often start browsing from pages identified by less precise but more easily constructed queries, instead of spending time to fully specify their search goals.[6] Aside from requiring additional time during the seeking processes, the burden on users is high and the benefits are not always clear, therefore the effectiveness of explicit techniques may be limited.

On the contrary, implicit feedback collects information about users while they perform their regular tasks. Basically, it unobtrusively draws usage data by tracking and monitoring user behavior, for example, by means of server access logs or query and browsing histories, e.g., see Refs. 4,7–11. Implicit and explicit feedback based on the same amount of information, namely snippets from result lists, are reasonably consistent in search en-

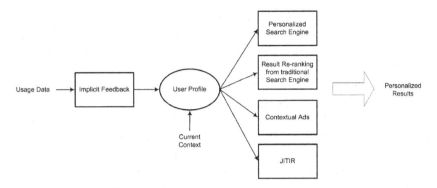

Fig. 2.1. A user profile based on implicit feedback techniques analyzes usage data, e.g. server access logs or query and browsing histories, and current working contexts in order to build representations of information needs of users.

gine domains. The fact that implicit feedback is readily available in large quantities overcomes possible bias in users' decisions, sometimes influenced by the trust they have in retrieval functions, and by the overall quality of result sets.[12]

Once enough knowledge has been collected, user profiles can be employed in search engines, to directly provide personalized results (see Fig. 2.1). For instance, Google Personalized Search records a trail of all queries and Web sites the user has selected from the results, building a profile according to that information. During the search activity, the search engine adapts the results according to needs of each user, assigning higher scores to the resources related to what the user has seen in the past. Alternatively, user profiles can take part in a distinct re-ranking phase. Some systems implement this approach on the client-side, where the software connects to a search engine, retrieving query results that are then analyzed locally, e.g. Refs. 13–15.

Just-in-Time IR (JITIR)[16] is a further interesting approach for personalized search where the information system proactively suggests information based on a person's working context. Basically, the system continuously monitors the user's interaction with the software, such as typing in a word processor or browsing the Web, in a nonintrusive manner, automatically identifying their information needs and retrieving useful documents without requiring any action by users (see also Watson[17] as a further instance of the JITIR approach). The same approach could be employed to provide contextual advertisements to users.

In this work, we propose a user modeling approach based on a cognitive theory of retrieval processes in human memory. The theory allows us to define not only suitable structures to organize concepts, but also how the information is stored and kept updated during the information seeking process. The user model is able to unobtrusively learn user needs by means of implicit feedback techniques. Unlike many traditional user models, the proposed approach takes under consideration the current user context in order to provide personalization related to the user's current activity. Moreover, the user model's learning phase does not need any off-line phase, during which the user provides data in order to train the model.

We also take into consideration an additional modeling approach based on Hyper-space Analogue to Language (HAL),[18] a semantic memory model that shares some similarities with the SAM-based model.

After introducing relevant related work in Sec. 2.2, Section 2.3.1 introduces SAM, the general theory on which one of the proposed user modeling approaches is based on, which is deeply discussed in Sec. 2.3.2. Section 2.3.3 presents the second approach for user profiling based on the HAL model. In what follows, i.e., Section 2.4, we discuss a wide evaluation of the two approaches in the context of Internet browsing activities. The last section concludes the chapter with final remarks and future works.

2.2. Related Work

Several statistical approaches for user profiling have been proposed in order to recognize user information needs while interacting with information sources. Nevertheless, a few of them are based on formal cognitive theories and implicit feedback techniques.[19]

A number of approaches that address the personalization task are based on the traditional content-based IR approach and explicit feedback techniques, e.g. Refs. 20–23. Some approaches reduce the burden on users to provide explicit feedback combining content and collaborative techniques.[24] Interesting approaches employ Naïve Bayes classifiers to build representations of the user needs, such as Syskill and Webert.[25] Personal WebWatcher[26] employs Bayes classifiers to represent the keywords contained in the hyperlinks selected by users, helping the user browse the Web highlighting interesting hyperlinks on visited Web pages. It is one of the first prototypes based on implicit feedback techniques.

Speretta and Gauch[15] analyze search histories in order to build user models based on ODP categories. Teevan et al. obtain interesting results performing a re-rank of search engine results according to user profiles built from the information created, copied, or viewed by a user, such as Web pages, emails, documents stored on the client PC, etc. Watson[17] monitors the user's actions and the files that he is currently working on to predict the users needs and offer them related resources. The TFxIDF technique[27] is used to create the contextual query based on the currently active window that is submitted to the information sources, i.e. search engines. These kinds of user models have many similarities with those we included in our evaluation.

Two systems use natural language processing and semantic or keyword networks in order to build long term user profiles and evaluate the relevance of text documents with respect to a profile. SiteIF project[28] uses semantic networks built from co-occurrence frequencies among keywords in News corpora, where each node represents the meaning of one keyword in given news, identified by means of the WordNet database.[29] ifWeb prototype[30] makes use of a network of keywords in order to create a representation of the available topics in one domain. The explicit feedback updates the user model adding or removing subsets of topics, i.e. subnetworks, judged interesting for the network associated with the user.

Two works based on cognitive theories are focused on the prediction of user actions, SNIF-ACT model[31] exploits ACT-R theory's concepts[32] such as the declarative and procedural knowledge, trying to represent all the items a user can deal with during the search, e.g. links, browser buttons, etc. and simulates users' actions during information-seeking processes by means of a set of production rules. The action selection considers the mutual relevance between the user goals and the current Web contents. Cognitive Walkthrough for the Web (CWW)[33] looks at the degree of similarity between user goals and heading/link texts by means of Latent Semantic Analysis, a technique that estimates the semantic relatedness of texts, based on a statistical analysis of a large corpus.[34] Even though they have not been employed in the personalization domain, the underlying cognitive models and techniques are in part related to our work.

2.3. SAM-based User Modeling Approach

We hereby provide a brief introduction of the SAM theory before investigating in depth the proposed user modeling approach.

2.3.1. *SAM: search of associative memory*

In this section, we give a short description of the SAM theory. For a closer examination of this theory, see, for example, Ref. 35.

SAM is a general theory of retrieval from long-term memory that considers both the structure of the memory system and the processes operating within it. The structure refers to the items represented and their organization in the memory system, the processes refer to the main activities that occur within the memory, such as learning and recall of the stored information.

The memory is organized in two parts: Long-Term Store (LTS) and Short-Term Store (STS). The STS shows two key features: a limited capacity and a proneness to "forget" its content (if the buffer size is reached, an item at random will be replaced). It can be regarded as a temporarily activated subset of information enclosed in the permanent LTS storage, which contains all prior information plus new information transferred from STS. The role of STS corresponds to a working space for control processes, such as coding, rehearsal, decision-making, etc. When a new external sensory input occurs, the related information is analyzed through the LTS structure, and data correlated with the input is activated and placed in the STS.

Both kinds of memories consist of unitized *images*, that is, objects that may be learned and recalled. Images also include temporal-contextual features that are not included in the discussion in order to make the description simple. The retrieval (or recall) process is based on the associative relationships between probe cues in STS, and LTS memory images. In the SAM theory, probe cues are the pieces of information the subject has with regard to the current task to be accomplished, e.g. information from a question, information retrieved earlier in the search, etc. The cue set activates a subset of LTS images. To what degree each of the images is activated is determined by a matrix that gives the strengths of relationship between each possible probe cue and each possible image (Fig. 2.2).

The recall process determines what images are sampled and made available to the user for evaluation and decision-making. It is convenient to organize the process in two phases:

- At each step of the process, STS cues are used as probes of LTS. The probability to *sample* a LTS's image is a function of the strength of association between the probe cues and the various images in LTS.
- The sampled LTS images are then accessed and evaluated by the so-

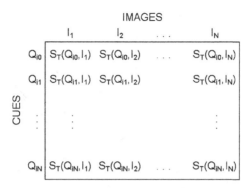

Fig. 2.2. The *strength matrix* used in the sampling and recovery processes to determine the probability of selection of words. It represents the information contained in LTS. The generic entry in the matrix corresponds to the strength between cue Q and word I (I stands for image in the SAM theory).

called *recovery* process. This process depends on the strength between the selected images and the probe cues. The recovered images will be stored in the STS.

The SAM Simulation (SAMS) approach is employed to build and keep the LTS structure updated. This approach consists of a buffer rehearsal process[36] that updates the image–image strength as a function of the total amount of time the pair of objects are simultaneously present in the STS. For example, let t_{ij} be the time that objects i and j are together in the STS simultaneously, then the strength is $S_T(i,j) = a\ t_{ij}$ where a is a parameter. The strength between a probe set and an image also increases when a successful recovery occurs.

2.3.2. *The user modeling approach*

During each information-seeking session, when users are looking for documents that satisfy their information needs, they have to assemble and deal with sets of concepts related to these needs. The identification of these concepts allows us to personalize the human-computer interaction, e.g. improving the ranking of resources retrieved by traditional tools such as search engines or recommending new resources to the user. For this reason, we have decided to ground our user modeling approach in the cognitive theory of human memory developed by Raaijmakers and Shiffrin.[35] It suggests important characteristics that have to be implemented in structures used to store memory concepts, and gives important recommendations about the

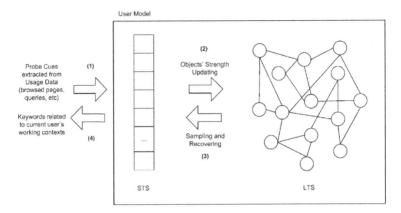

Fig. 2.3. The user profile is organized in two memory structures: STS and LTS. The former is a working buffer that contains probe cues extracted from usage data and subsets of LTS images (or words). The latter is the permanent store, containing all prior information.

algorithms to store and retrieve information from these structures.

Figure 2.3 shows the user model's internal organization and how it interacts from the outside. Probe cues are drawn from usage data (1), which is information related to users' behavior while interacting with the system. Examples of usage data are past queries submitted to search engines or contents of the browsed pages. This information is transferred in the temporary STS store (2) and is used by the sampling and recovering processes to probe the LTS (3), the permanent store containing all prior information transferred from STS. The probability to activate and transfer information from LTS to STS is a function of the strength of association between the STS probe cues and the information stored in LTS. These strength values saved in the LTS structure are assumed to be proportional to the total amount of time a given information unit remains in STS. The successfully recovered LTS units are transferred to STS and correspond with the information related to the user's current working context, e.g. a browsed page or an e-mail the user is currently writing. This information is useful to personalize the interaction with the system (4), for example, generating queries to several sources of information and presents them in a separate window.

2.3.2.1. *LTS and STS*

The SAM theory does not require particular memory representations for LTS and STS buffers, therefore we have to define the information encoding, e.g. the kind of images stored in the LTS and the STS cues used for probing. In our approach, both kinds of memories consist of words, therefore they correspond with the unitized information that may be learned and recalled. This choice is justified by its simplicity and adaptation to the Web personalization context used in the evaluation. We have considered an environment where documents are represented by text sentences, therefore words seem the obvious elements that could be stored and analyzed.

No temporal or contextual information has been stored along with each word in this prototype. Nevertheless, Natural Language Processing (NLP) techniques can be successfully employed in order to assign a unique semantic meaning to each word extracted from usage data, for example, analyzing the sentence in which the word occurs and using this co-occurrence information to query the *WordNet* database.[29]

LTS stores associative relationships between words. The amount of word–word information stored is assumed to be proportional to the total amount of time those two words are simultaneously present in the rehearsal buffer STS. The associations are stored in a *strength matrix* (see Fig. 2.2). Given a cue Q and a word I, the matrix stores the strength between these two objects $S_T(Q, I)$.

The STS size is a parameter of the system. Miller claimed that about seven chunks could be held into this kind of short-memory,[37] where a *chunk* stands for an integrated unit of information. Instead of remembering a single stimulus, humans usually group input events, apply a new name to the group, and then remember the new name rather than the original input events. In this way, since the memory span is a fixed number of chunks, it is possible to increase the number of bits of information that it contains simply by building larger and larger chunks by means of a *recoding* process. Nevertheless, this kind of process has not been included in the user modeling approach.

The choice to use words as the unitized information that may be learned and recalled from the memory, and the absence of the recoding process lead us to reconsider the initial STS size. During empirical evaluations, users were asked to remember as many words as they could after having read sequences of documents. Being able to organize many words into single chunks, they could recall many more words than expected. Following these preliminary evaluations, we decided to set the STS size in the 25-100 unit range (see Sec. 2.4.5 for experimental results).

2.3.2.2. Sampling and Recovery

As explained in the previous section, the retrieval process consists of two phases: the sampling of words in the LTS, and their evaluation in a recovery process. Both steps exploit the strength matrix, which corresponds to the information stored in the LTS.

Given the current cues Qs, the sampling phase draws the probabilities for each word I_i in LTS as a function of the strength of association between them:

$$P_S(I_i|Q_1, Q_2 \cdots Q_M) = \frac{\prod_{j=1}^{M} S_T(Q_j, I_i)^{W_j}}{\sum_{k=1}^{N} \prod_{j=1}^{M} S_T(Q_j, I_k)^{W_j}} \quad (2.1)$$

where M is the current number of cues in STS, and N is the matrix dimension.

Equation (2.1) assigns high probabilities to words with the highest product of strengths, hence, those that tend to be greatly associated with all the current cues. W_j represents weights used to give different importance to the cues. For instance, in the Web domain, Inverse Document Frequency (IDF) values[38] can decrease the weight for words that frequently occur in a given corpus, therefore very common and hardly relevant. In our evaluation W_j takes on the expression:

$$W_j = \left(\frac{\text{idf}_j}{\max \text{Idf}}\right)^3 \quad (2.2)$$

where $\text{idf}_j = \log \frac{|\text{Docs}|}{df_j}$, df_j is the Document Frequency of the j word, $\max \text{Idf} = \log |\text{Docs}|$ and Docs is a given input set of documents.

Once a word has been sampled, the recovery process takes place. The probability to successfully retrieve the sampled word I_i corresponds to:

$$P_R(I_i|Q_1, Q_2 \cdots Q_M) = \frac{\sum_{j=1}^{M} S_T(Q_j, I_i)^{W_j}}{\max_{k \in [1,N]} \sum_{j=1}^{M} S_T(Q_j, I_k)^{W_j}}. \quad (2.3)$$

This expression differs considerably from others in the literature, e.g. Ref. 33:

$$P_R(I_i|Q_1, Q_2 \cdots Q_M) = 1 - e^{[-\sum_{j=1}^{M} W_j S_T(Q_j, I_i)]}. \quad (2.4)$$

The reason concerns the size of the rehearsal buffer, which is larger than the size used in other systems. If there are many cues, the exponent in Eq. (2.4) gets very low values and the probability tends to 1. In other words, the cue effect is so marginal that all the sampled images will be successfully recovered. The normalization in Eq. (2.3) prevents this effect. As in the original formulation, large cue weights affect the recovery positively and probabilities get high values even if one strength is high.

2.3.2.3. Learning

The strength matrix is kept updated according to the total time a word or a pair of words are stored together in the STS buffer. Given t_i the time spent in the buffer by the word I_i, and t_{ij} the time images I_i and I_j occur together in the buffer, we have:

$$S_T(I_i, I_j) = S_T(I_j, I_i) = b\, t_{ij}, \quad t_{ij} \neq 0,$$
$$S_T(I_i, I_i) = c\, t_i.$$

If a word pair has never appeared together in the buffer, they assume a nonnegligible residual strength d. The values b, c and d are parameters of the model.

The strengths are also increased whenever a successful recovery occurs. In this case, the strength between the cues Q_is and the word I_j, and the self-association strength $S_T(I_j, I_j)$ is incremented:

$$S'_T(Q_i, I_j) = S_T(Q_i, I_j) + f, \quad t_{ij} \neq 0$$
$$S'_T(I_j, I_j) = S_T(I_j, I_j) + g$$

where S is the strength before incrementing, and f and g are further parameters of the model.

The currently activated words in STS and their relationships with the words in LTS are used to build new LTS relationships. An important feature of the model is that units of information in the STS tend to be stored jointly. For example, when the user is browsing a given page and some content is extracted and given as input to the model, the words are stored in LTS with relationships that connect the words to each other. When some of these words are recovered, the relationships help to retrieve the correlated information.

In natural language domains, such as the Web, this kind of implicit context helps to disambiguate the meaning of words. The relationships among them connect each word with the implicit context present at the time of learning. It may happen that users analyze the same word again, but in different contexts. In this case, different sets of connections are built between the word and the new context. Probing the LTS with some words related to the right context help recognize the right meaning of the word.

Moreover, the word significance depends on the user's current working context. Traditional approaches assign weights to words according to explicit and/or implicit feedbacks without taking under consideration the user's current activity. In our approach, a word is judged interesting if it is connected to the context, that is, the information stored in the control and decision-making buffer STS.

2.3.2.4. *Interaction with Information Sources*

After having explained the sampling, recovering and learning processes, we briefly describe the general retrieval process that occurs each time the user interacts with information sources. For a detailed description, see Ref. 39.

The process can be broken down to two inner retrieval processes that are based on the SAM cognitive model. The first process makes K_{MAX} attempts to sample and recover words stored in LTS by means of the current probe cues (the context stored in STS). If a word I_i has been successfully sampled, the strengths $S_T(Q_j, I_i)$ with the current STS's words Q_js, and the self-association $S_T(I_i, I_i)$ are increased, and the word is included in the STS. The updated buffer will be used to sample other correlated words. A further retrieval process corresponds to a *rechecking* phase. It ensures that all associative retrieval routes starting from the STS's words have been checked thoroughly, at least L_{MAX} times.

The main retrieval process concerns the human-computer interaction with information sources. In our domain, we analyze the user's current activity and extract content that is given to the model as probe cues. Examples of probe cues are: queries, document's snippets, categories selected by the user. The available information is included in the current context stored in STS, and the retrieval process takes place to recover and suggest correlated information and, at the same time, update the LTS. The process periodically checks if the current user activity is changed, e.g. the user closed the browser window and opened a word processor. In this case, the temporary information stored during the session is wiped out and the STS is cleared.

Each time the recovery process is completed, the STS words correspond to the information related to the current information needs of the user. As described in the evaluation they can be used for query expansion, enhancing the query using words or phrases related to the set of documents seen by the user and the current user activity.

Traditional interactive systems ask users to mark documents as relevant or nonrelevant. These documents are used as training data for a relevance feedback query expansion approach. On the contrary, the proposed user modeling approach is based on implicit feedback techniques, where usage data are analyzed in order to draw information used to build the model and keep it updated.

The described user modeling is based only on an additive learning process. At the same time, it is always possible to ignore concepts judged not interesting. In fact, the learning process tends to increase the relationship strengths between concepts that frequently occurred during the interaction with information sources. Wrong concepts may be included in the model, but their strengths with other words will not be increased if they no longer appear in the probe cues. Therefore, the probabilities to recover these concepts decrease as new concepts are included in the model. In other words, "forgetting" follows the failure in the attempt to retrieve concepts.

2.3.3. HAL-based User Modeling Approach

The Hyperspace Analogue to Language (HAL) model[18] of memory grounds its roots in psychological theories of word meaning. In these theories the meaning of words learnt by users is a function of the contexts in which that word has occurred in. The word meaning of *Jaguar* for example could be related to an animal, a car manufacturer, the operating system, etc.. The context of the word, that is, the words the co-occour with it, determines the particular concept the given peace of text intends to communicate. Humans are able to use the context neighborhoods to match words with similar co-occurring items and to derive a specific word from its neighborhood.[40]

In computational terms, the HAL model represents each word as a vector of weights corresponding to other words. Given a corpus of textual documents, by utilizing a global co-occurrence learning algorithm, HAL stores weighted co-occurrence information forming a matrix of co-occurrences for words in the context of other words. Two vectors can then be compared to each other, giving an overall measure of the similarity of two words' co-occurrence patterns, with the underlying hypothesis that two words that have occurred in similar contexts will have similar meanings.

The weighing scheme of the HAL vectors is based on a sliding window of fixed length, typically 8-10 words, that runs across the text, calculating the word co-occurrences values along the way. The HAL model is represented with a $N_h \times N_h$ matrix of integers, where N_h is the number of distinct words appearing in text. Given two words, whose distance within the window is d, the weight between them is determined by $w - d + 1$, where w is the window size. This basically means that closer words get higher scored than words farther away.

Table 2.1. Sample HAL matrix for the sentence "What goes around comes around" using a window of 5 words.

	around	comes	goes	what
around	4	5	8	6
comes	5	0	4	3
goes	0	0	0	5
what	0	3	0	0

The HAL model has been exploited in a broad range of cognitive and data analysis tasks, such as proper name semantics, semantic and grammatical categorization, simulating memory in mental disorders, semantic constraints during syntactic processing, etc. (e.g., Refs 41–45).

In our context, the HAL model represents the concepts one user has identified and analyzed during the interaction with information sources. In other words, documents browsed or edited become the corpus on which the HAL model is trained. Once the hyperspace of concepts has been formed, given a keyword it is possible to analyze its context, i.e., the lexical co-occurring words, by the vector of the HAL matrix. The words with higher weights can be employed for IR personalization by query expansion techniques, or to filter incoming streams of information.

In our evaluation, the Inverse Vector Frequency (IVS) function[46] re-weights the HAL values according to the measure of informativeness of words:

$$IVF(w) = \frac{\log(\frac{N_h + 0.5}{n})}{\log(N + 1)} \tag{2.5}$$

as the Inverse Document Frequency (IDF) does likewise in the IR domain; n is the number of elements of the vector of the word w with a value greater than 0.

Conceptually, HAL models have some sort of similarities with the SAM-based approach. The strength matrix in SAM stores how long two given information units remains in the STS. The HAL matrix counts the times two words were close together. If we consider words as units of information stored in the LTS and STS of SAM, ignoring any additional temporal or contextual information, the two models' structures coincide. As for the training phase, the HAL sliding window of 8-10 words is not comparable to 50-100 words of the STS. Moreover, the First-In First-Out policy of the sliding window differs from the random deletion that happens when the STS buffer size is reached. Therefore, the two matrices store data that we are not able to compare. Lastly, the recovery process described in the previous sections used to retrieve related words used for query expansion is composed of two distinct aleatory process, while HAL uses a statistical selection based on co-occurrence frequencies.

2.4. Evaluation

In this section, we discuss several experiments to determine the performance of the SAM-based and HAL-based user modeling approaches in Web browsing tasks.

A traditional IR-based approach provides us the benchmark to compare our evaluation results. It is based on a content-based technique where documents are represented through the Vector Space model (VSM). A Relevance Feedback (RF) technique updates the model according to the content of the training pages. The keywords with highest rank, measured by means of the TFxIDF measure, are used for query expansion.

2.4.1. *Evaluating User Models in Browsing Activities*

The amount of the available information that can be exploited during any information-seeking task makes the Web a fundamental environment where user modeling approaches can be evaluated. Recognizing what users are looking for during browsing sessions, that is, their information needs, is the first step towards efficient personalization techniques.

In order to employ our user modeling approach on the Web, we must resort to a methodology to identify the probe cues, i.e., information related to the current needs, used to train the user models. The notion of *information scent*[47,48] developed in the context of Information foraging theory[49], already evaluated in different tasks,[47] is a valid approach in recognizing these cues by means of the anchor text associated to each link. While browsing, users use these *proximal cues* to decide whether to access the distal content, that is, the page pointed by the link. Formally, the information scent is the imperfect, subjective perception of the value or cost of the information sources obtained from proximal cues.

Each time the user selects a link and visits the corresponding page, the link's anchor text can be extracted and used in learning and recall processes. However, preliminary evaluations show how sometimes this information is not enough to recognize valuable probe cues, especially if the text consists only of a few common words, e.g. "full story", "page two", "link", "rights reserved", etc.

For this reason, we have developed an algorithm to collect information related to a given link selection: the anchor text is combined with the title of page visited by the user. Afterwards, by means of the page's Document Object Model,[a] the page is divided into units whose boundaries are defined by a subset of HTML tags, e.g. TR, P, UL, etc. and the text of the deepest unit that contains the link is retrieved. Finally, the retrieved text plus anchor and title are compared to the other units of the pointed page, in order to find further related text. The text comparison is based on the IR similarity function used in the TextTiling algorithm.[50] For a deeper analysis of the cue-extraction technique, see Ref. 51.

As for the SAM-based user modeling, in this evaluation no temporal or contextual information has been stored along with each word, nor have NLP techniques to analyze semantic meanings been included in this first prototype. After a preliminary evaluation, we have chosen the following values for the SAM parameters: $a = 0.1$, $b = 0.1$, $c = 0.1$, $d = 0.2$, $e = 0.7$, $f = 0.7$ and $g = 0.7$. As for the HAL model, the sliding window is set to 10 words, according to past experiments.[40,52]

2.4.2. Corpus-based evaluation

After a preliminary evaluation based on human subjects and subsets of browsers' histories,[53] we have developed a more formal and accurate framework that simulates implicit feedback techniques for browsing activities.[39] Instead of making *ad hoc* corpus of documents based on Web browsing sessions, and manually recognizing the information needs that led users through specific paths, we decided to consider some categories in a directory service Web site for training and test phases. The paramount advantage is that measures of search effectiveness, e.g. precision and recall, can be adopted without considering time consuming and costly experiments with human subjects.

After having randomly chosen N_{odp} categories from the Open Directory

---
[a]http://www.w3.org/DOM/

Project[b] (ODP), a subset of the entries in each category is selected for the training phase. Each category corresponds to a different user information need. For example, in the third level of the OPD category we could have:

(i) BUSINESS, ENERGY AND ENVIRONMENT, MANAGEMENT
(ii) HEALTH, OCCUPATIONAL HEALTH AND SAFETY, CONSULTANTS
(iii) SHOPPING, HOME AND GARDEN, GARDEN SHOPS

In order to provide usage data for training the models, we employed two different methodologies. The former described in Sec.2.4.1 is based on anchor text and related text retrieved from pointed Web pages. In our evaluation, the first page is built by the information extracted from the ODP category, e.g, category name, title e description of the entry in the ODP category, while the second page corresponds to the content of the entry page. The second methodology simply draws the text related to the user needs extracting the whole pages' content. We expect that the latter brings extra information loosely related to the current needs, because of the Web page elements, such as advertisements, browsing support items, etc.

The evaluation is based on the user model's ability to suggest further Web sites related to the user needs, namely, the remaining Web site entries that are not used for training in the chosen ODP categories. In particular the cross validation sets are 25% for training and 75% testing. In general, users analyze a small number of pages on a given topic of interest, while the available pages of interest are relatively large sets. This justifies the chosen ratio.

Basically, the name of the ODP category is given as probe cues and the related keywords returned by the user models are used for query expansion. A standard IR search engine indexes the Web sites contained in the first three layers of the ODP hierarchy . The following cosine rule is used to match the expanded query Q with the ODP entries Ds:

$$M(Q,D) = \sum_{t \in Q} \frac{tf_{Q,t} \cdot \text{idf}_t}{\sqrt{\sum_{t \in Q}(tf_{Q,t} \cdot \text{idf}_t)^2}} \cdot \frac{tf_{D,t} \cdot \text{idf}_t}{\sqrt{L_D}} \text{boost}_t \qquad (2.6)$$

where t is a term of the query Q, L_D is the number of terms in the document to analyze D, $t_{x,t}$ is the term frequency of t in the query or document x, and finally idf_t is the inverse document frequency of t. The expanded queries are composed of the category name and the suggested keywords.

[b]http://www.dmoz.org

The precision is the measure employed to estimate the performance of the different models. For each query, i.e., ODP topic, the test set is compared to the list of the top 50 results retrieved during the query expansion. The precision is the number of results that are in the test set divided by the size of the test set.

The different evaluation phases are summarized as follows:

(a) Random selection of N_{odp} ODP categories corresponding to stable information needs.
(b) Random selection of 25% urls in each category for the training phase.
(c) Extraction of probe cues by means of the algorithm described in Sec. 2.4.1.
(d) Training of user models:

 (1) SAM: learning and recovery processes, see Sec. 2.3.2.2
 (2) HAL: sliding window, see Sec. 2.3.3
 (3) VSM: relevance feedback

(e) The category names and descriptions are given as probe cues to the UM in order to retrieve related sets of keywords.
(f) Query expansion with the retrieved keywords.
(g) Precision of the top 50 documents returned by a search engine which indexed the ODP urls.

In the rest of this section we discuss the performance of the modeling approaches as a function of different features considered. In particular, we begin analyzing how the number of topics of interest affects the precision of the models.

2.4.3. Precision vs. Number of Topics

The first evaluation regards the difference in performance levels as a function of the number of ODP categories, i.e, sessions that correspond to distinct user needs, which the models analyze. User models should be able to keep several and typically not correlated needs for filtering and recommending tasks. Good models are able to retrieve concepts and keywords related to the subset of needs the user is currently showing interest in.

Figure 2.4 shows the precision of the three models: one based on VSM and a traditional relevance feedback technique, and the other two approaches based on HAL modeling and the SAM theory. On average the RF-based model shows low performance. This traditional modeling approach does not take into consideration the current context of users' ac-

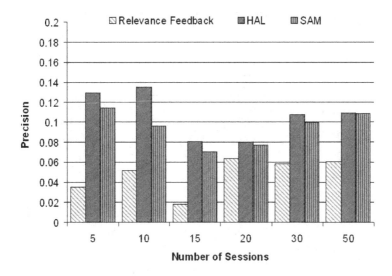

Fig. 2.4. A chart of the precision values obtained in the evaluation as a function of the ODP categories (N_{odp}) used for training and test.

tivities, i.e., probe cues extracted during the browsing. For this reason, its personalization does not depend on the particular query, and the expansion considers keywords related to many different topics at the same time.

Table 2.2. Precision values as a function of N_{odp} categories

Sessions	VSM-RF	HAL	SAM
5	0.035	0.129	0.114
10	0.051	0.134	0.096
15	0.018	0.080	0.070
20	0.063	0.080	0.077
30	0.058	0.107	0.100
50	0.060	0.109	0.108

The HAL-based modeling shows slightly better performance in comparison with the SAM approach, even though after a certain number of sessions, the performance difference is not noticeable. Probably the SAM approach has better chances to store relationships between concepts coming from different sessions, while HAL might include ambiguous correlations if the number of topics gets high levels.

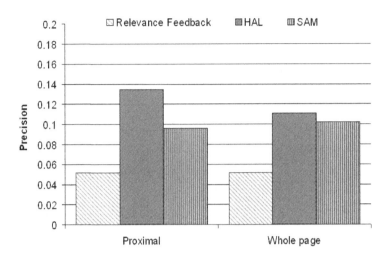

Fig. 2.5. Precision results when two different cue extraction techniques are employed.

The trend among the number of sessions depends on the page content used for training. Between 15 and 20 sessions, the evaluation included several topics from which it was hard to draw relevant data to train the models.

In general, the absolute precision values are low (less than 0.15, as shown in Tab. 2.2). As a natural consequence of the chosen domain, we have many Web pages with no content (only images or animations), or home-pages that have short content related to the main topic. If we consider a corpora of roughly 200.000 resources, as in our case, it is not easy for simple content-based query expansions to obtain high precision levels.

2.4.4. *Precision vs. Extracted Cues*

In the second evaluation we simulate how the models are able to deal with usage data that contain noisy information. Instead of employing the extraction technique based on proximal cues, see Sec.2.4.1, we have considered the whole page contents during the learning phase.

Taking 10 ODP categories for training, we can see in Fig. 2.5 that when the whole page content is considered, the difference between the SAM and HAL modeling is lower. SAM-based models show higher tolerance to usage data with noise information. This is an advantage in domains where cue

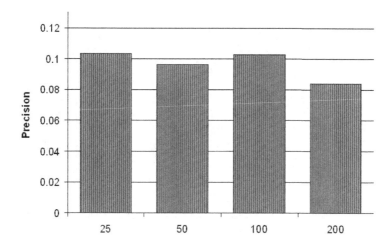

Fig. 2.6. Precision of the SAM model as a function of the STS size.

extraction techniques are hard to develop.

2.4.5. Precision vs. Size of STS

In the past experiments we set to 50 the size of the SAM STS. We performed the same experiments with different sizes of the STS. In particular, Fig. 2.6 shows the performance of SAM modeling for STS of 25, 50, 100 and 200 elements. It is possible to say that a too large size of STS negatively affects the learning, causing the inclusion of LTS relationships that degrade the global precision. A size between 25 and 100 shows the best results.

2.4.6. Precision vs. Number of Recovery Attempts

In the last evaluation, we try to recover additional keywords from the LTS running multiple attempts (for a detailed description of the SAM recovery process see Sec.2.3.2.2 and Ref. 39). Figure 2.7 shows how values greater than two do not increase the precision of the retrieval process. In other words, when we try to run multiple recoveries, the process draws keywords that are not related to the current needs of the user.

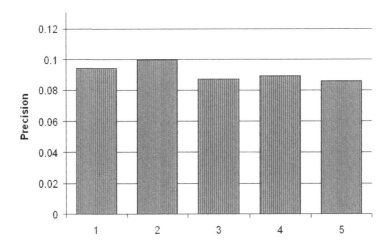

Fig. 2.7. Influence of the number of recovery attempts on SAM precision.

2.5. Conclusions

As a first step towards better modeling of user needs and interests, we have proposed two approaches based on cognitive theories and models. The goal is to represents basic human memory processes of learning and retrieval during the interaction with information sources.

To assess the performance of the two approaches, we have devised an evaluation framework based on the ODP category and standard IR measures, i.e., precision of result sets. The evaluation proves the good performance of SAM-based user modeling in providing keywords related to the current Web browsing needs.

If the users are interested in a few topics or the usage data exploited during model learning is particular noisy, HAL-based models evaluated in the same context obtain slightly higher degree of precision. In other words, the STS and the related learning process of the SAM-based modeling do not provide any significant advantage.

As the proposed approaches deals with low levels human memory processes, it is possible to imagine more complex models of the human-computer interaction with information sources. These models represent information-seeking strategies and plans users undertake when a particular task ought to be accomplished. Our future work will be towards this high-level modeling.

Further potential enhancements concern the inclusion of contextual information in each word stored in STS and LTS. This kind of information is available through common categorization techniques that assign a category to each analyzed document. NLP techniques able to assign unique semantic meanings to each word can increase the performance of the approach further.

References

1. R. White, J. M. Jose, and I. Ruthven. Comparing explicit and implicit feedback techniques for web retrieval: Trec-10 interactive track report. In *TREC*, (2001).
2. A. Wærn, User involvement in automatic filtering: An experimental study, *User Modeling and User-Adapted Interaction.* **14**(2-3), 201–237, (2004).
3. P. Anick. Using terminological feedback for web search refinement: a log-based study. In *SIGIR '03: Proceedings of the 26th annual international ACM SIGIR conference on Research and development in informaion retrieval*, pp. 88–95, New York, NY, USA, (2003). ACM Press. ISBN 1-58113-646-3. doi: http://doi.acm.org/10.1145/860435.860453.
4. J. Teevan, S. T. Dumais, and E. Horvitz. Personalizing search via automated analysis of interests and activities. In *SIGIR '05: Proceedings of the 28th annual international ACM SIGIR conference on Research and development in information retrieval*, pp. 449–456, New York, NY, USA, (2005). ACM Press. ISBN 1-59593-034-5. doi: http://doi.acm.org/10.1145/1076034.1076111.
5. J. Koenemann and N. J. Belkin. A case for interaction: a study of interactive information retrieval behavior and effectiveness. In *CHI '96: Proceedings of the SIGCHI conference on Human factors in computing systems*, pp. 205–212, New York, NY, USA, (1996). ACM Press. ISBN 0-89791-777-4. doi: http://doi.acm.org/10.1145/238386.238487.
6. J. Teevan, C. Alvarado, M. S. Ackerman, and D. R. Karger. The perfect search engine is not enough: a study of orienteering behavior in directed search. In *CHI '04: Proceedings of the SIGCHI conference on Human factors in computing systems*, pp. 415–422, New York, NY, USA, (2004). ACM Press. ISBN 1-58113-702-8. doi: http://doi.acm.org/10.1145/985692.985745.
7. P. K. Chan. Constructing web user profiles: A non-invasive learning approach. In *WEBKDD '99: Revised Papers from the International Workshop on Web Usage Analysis and User Profiling*, pp. 39–55, London, UK, (2000). Springer-Verlag. ISBN 3-540-67818-2.
8. M. Claypool, P. Le, M. Wased, and D. Brown. Implicit interest indicators. In *IUI '01: Proceedings of the 6th international conference on Intelligent user interfaces*, pp. 33–40, New York, NY, USA, (2001). ACM Press. ISBN 1-58113-325-1. doi: http://doi.acm.org/10.1145/359784.359836.
9. D. Kelly and J. Teevan, Implicit feedback for inferring user preference: a bibliography, *SIGIR Forum.* **37**(2), 18–28, (2003). ISSN 0163-5840. doi: http:

//doi.acm.org/10.1145/959258.959260.
10. F. Radlinski and T. Joachims. Query chains: learning to rank from implicit feedback. In *KDD '05: Proceeding of the eleventh ACM SIGKDD international conference on Knowledge discovery in data mining*, pp. 239–248, New York, NY, USA, (2005). ACM Press. ISBN 1-59593-135-X. doi: http://doi.acm.org/10.1145/1081870.1081899.
11. R. White, I. Ruthven, and J. M. Jose. The use of implicit evidence for relevance feedback in web retrieval. In eds. F. Crestani, M. Girolami, and C. J. van Rijsbergen, *Advances in Information Retrieval, 24th BCS-IRSG European Colloquium on IR Research Glasgow, UK, March 25-27, 2002 Proceedings*, vol. 2291, *Lecture Notes in Computer Science*, pp. 93–109. Springer, (2002).
12. T. Joachims, L. Granka, B. Pan, H. Hembrooke, and G. Gay. Accurately interpreting clickthrough data as implicit feedback. In *SIGIR '05: Proceedings of the 28th annual international ACM SIGIR conference on Research and development in information retrieval*, pp. 154–161, New York, NY, USA, (2005). ACM Press. ISBN 1-59593-034-5. doi: http://doi.acm.org/10.1145/1076034.1076063.
13. J. Pitkow, H. Schütze, T. Cass, R. Cooley, D. Turnbull, A. Edmonds, E. Adar, and T. Breuel, Personalized search, *Commun. ACM*. **45**(9), 50–55, (2002). ISSN 0001-0782. doi: http://doi.acm.org/10.1145/567498.567526.
14. A. Micarelli and F. Sciarrone, Anatomy and empirical evaluation of an adaptive web-based information filtering system, *User Modeling and User-Adapted Interaction*. **14**(2-3), 159–200, (2004). ISSN 0924-1868. doi: http://dx.doi.org/10.1023/B:USER.0000028981.43614.94.
15. M. Speretta and S. Gauch. Personalized search based on user search histories. In *Web Intelligence (WI2005)*, France, (2005). IEEE Computer Society.
16. B. J. Rhodes. *Just-In-Time Information Retrieval*. PhD thesis, MIT Media Laboratory, Cambridge, MA (May, 2000). URL citeseer.ist.psu.edu/rhodes00justtime.html.
17. J. Budzik, K. J. Hammond, and L. Birnbaum, Information access in context., *Knowledge-Based Systems*. **14**(1-2), 37–53, (2001).
18. K. Lund and C. Burgess, Producing high-dimensional semantic spaces from lexical co-occurrence, *Behavior Research Methods, Instruments and Computers*. **28**(2), 203–208, (1996).
19. A. Micarelli, F. Gasparetti, F. Sciarrone, and S. Gauch. Personalized search on the world wide web. In eds. P. Brusilovsky, A. Kobsa, and W. Nejdl, *The Adaptive Web: Methods and Strategies of Web Personalization*, vol. 4321, *Lecture Notes in Computer Science*, pp. 195–230. Springer Berlin, Heidelberg, Berlin, Heidelberg, and New York, (2007). ISBN 3-540-72078-2. doi: http://dx.doi.org/10.1007/978-3-540-72079-9_6.
20. T. Joachims, D. Freitag, and T. M. Mitchell. Webwatcher: A tour guide for the world wide web. In *Proceedings of the 15h International Conference on Artificial Intelligence (IJCAI1997)*, pp. 770–777, (1997).
21. A. Moukas and P. Maes, Amalthaea: An evolving multi-agent information filtering and discovery system for the WWW, *Autonomous Agents and Multi-*

Agent Systems. **1**(1), 59–88, (1998).
22. H. Lieberman, N. W. V. Dyke, and A. S. Vivacqua. Let's browse: a collaborative web browsing agent. In *Proceedings of the 4th International Conference on Intelligent User Interfaces (IUI'99)*, pp. 65–68, Los Angeles, CA, USA, (1998). ACM Press.
23. M. Montaner, B. López, and J. L. D. L. Rosa, A taxonomy of recommender agents on the internet, *Artificial Intelligence Review.* **19**(4), 285–330 (June, 2003).
24. M. Balabanovic and Y. Shoham, Fab: content-based, collaborative recommendation, *Communications of the ACM.* **40**(3), 66–72, (1997). ISSN 0001-0782.
25. M. J. Pazzani, J. Muramatsu, and D. Billsus. Syskill webert: Identifying interesting web sites. In *Proceedings of the National Conference on Artificial Intelligence (AAAI-96)*, pp. 54–61, Portland, OR, USA (August, 1996). AAAI Press.
26. D. Mladenic. Using text learning to help web browsing. In *Proceedings of the Ninth International Conference on Human-Computer Interaction*, (2001).
27. G. Salton, A. Wong, and C. Yang, A vector space model for automatic indexing, *Commun. ACM.* **18**(11), 613–620, (1975). ISSN 0001-0782. doi: http://doi.acm.org/10.1145/361219.361220.
28. B. Magnini and C. Strapparava, User modelling for news web sites with word sense based techniques, *User Modeling and User-Adapted Interaction.* **14**(2), 239–257, (2004).
29. G. A. Miller and C. Fellbaum. Lexical and conceptual semantics. In eds. B. Levin and S. Pinker, *Advances in Neural Information Processing Systems*, pp. 197–229. Blackwell, Cambridge and Oxford, England, (1993).
30. F. A. Asnicar and C. Tasso. ifweb: a prototype of user model-based intelligent agent for document filtering and navigation in the world wide web. In *Proceedings of Workshop Adaptive Systems and User Modeling on the World Wide Web (UM97)*, pp. 3–12, Sardinia, Italy, (1997).
31. P. Pirolli and W.-T. Fu. Snif-act: A model of information foraging on the world wide web. In *Proceedings of the 9th International Conference on User Modeling*, Johnstown, PA, USA, (2003).
32. J. R. Anderson and C. Lebiere, *The atomic components of thought.* (Lawrence Erlbaum Associates, Mahwah, NJ, U.S.A., 2000).
33. M. H. Blackmon, P. G. Polson, M. Kitajima, and C. Lewis. Cognitive walkthrough for the web. In *Proceedings of the ACM conference on human factors in computing systems (CHI2002)*, pp. 463–470, Minneapolis, Minnesota, USA, (2003).
34. T. K. Landauer and S. T. Dumais, A solution to platos problem: The latent semantic analysis theory of acquisition, induction, and representation of knowledge, *Psychological Review.* **104**, 211–240, (1997).
35. J. G. Raaijmakers and R. M. Shiffrin, Search of associative memory, *Psychological Review.* **88**(2), 93–134 (March, 1981).
36. R. C. Atkinson and R. M. Shiffrin. Human memory: A proposed system and its control processes. In eds. K. Spence and J. Spence, *The psychology of*

learning and motivation: Advances in research and theory, vol. 2, pp. 89–195. Academic Press, New York, N.Y., U.S.A., (1968).
37. G. A. Miller, The magical number seven, plus or minus two: Some limits on our capacity for processing information, *Psychological Review.* **63**, 81–97, (1956).
38. K. S. Jones, Idf term weighting and ir research lessons, *Journal of Documentation.* **60**(5), 521–523, (2004).
39. F. Gasparetti and A. Micarelli, Personalized search based on a memory retrieval theory, *International Journal of Pattern Recognition and Artificial Intelligence (IJPRAI): Special Issue on Personalization Techniques for Recommender Systems and Intelligent User Interfaces.* **21**(2), 207–224 (March, 2007). ISSN 0218-0014.
40. C. Burgess, K. Livesay, and K. Lund, Explorations in context space: Words, sentences, discourse, *Discourse Processes.* **25**(2–3), 211–257, (1998).
41. J. Alison and C. Burgess, Effects of chronic non-clinical depression on the use of positive and negative words in language contexts, *Brain and Cognition.* **53**(2), 125–128, (2003).
42. C. Burgess and K. Lund. Modeling cerebral asymmetries in high-dimensional semantic space. In eds. M. Beeman and C. Chiarello, *Right Hemisphere Language Comprehension: Perspectives from Cognitive Neuroscience*, pp. 245–254. Lawrence Erlbaum Associates, Inc, Hillsdale, NJ, U.S.A., (1998).
43. R. A. Atchleya, J. Storya, and L. Buchananb, Exploring the contribution of the cerebral hemispheres to language comprehension deficits in adults with developmental language disorder, *Brain and Cognition.* **46**(1-2), 16–20, (2001).
44. C. Burgess, From simple associations to the building blocks of language: Modeling meaning in memory with the hal model, *Behavior Research Methods, Instruments and Computers.* **30**(2), 188–198, (1998). ISSN 0743-3808.
45. P. Conley, C. Burgess, and G. Glosser, Age vs alzheimer's: A computational model of changes in representation, *Brain and Cognition.* **46**(1-2), 86–90, (2001).
46. D. Song, P. D. Bruza, and R. Cole. Concept learning and information inferencing on a high-dimensional semantic space. In *ACM SIGIR 2004 Workshop on Mathematical/Formal Methods in Information Retrieval (MF/IR'2004)* (July, 2004). URL www.kmi.open.ac.uk/people/dawei/papers/mfir2004.pdf.
47. E. H. Chi. Transient user profiling. In *Proceedings of the Workshop on User Profiling (CHI2004)*, pp. 521–523, Vienna, Austria, (2004).
48. E. H. Chi, P. Pirolli, K. Chen, and J. Pitkow. Using information scent to model user information needs and actions on the web. In *Proceedings of the ACM Conference on Human Factors in Computing Systems (CHI2001)*, pp. 490–497, Seattle, WA, USA, (2001).
49. P. Pirolli and S. K. Card, Information foraging, *Psychological Review.* **106**, 643–675, (1999).
50. M. A. Hearst, Texttiling: Segmenting text into multi-paragraph subtopic passages, *Computational Linguistics.* **23**(1), 33–64, (1997).

51. F. Gasparetti and A. Micarelli. Exploiting web browsing histories to identify user needs. In *IUI '07: Proceedings of the 12th international conference on Intelligent user interfaces*, pp. 325–328, New York, NY, USA, (2007). ACM Press. ISBN 1-59593-481-2. doi: http://doi.acm.org/10.1145/1216295.1216358.
52. P. D. Bruza and D. Song. Inferring query models by computing information flow. In *CIKM '02: Proceedings of the eleventh international conference on Information and knowledge management*, pp. 260–269, New York, NY, USA, (2002). ACM Press. ISBN 1-58113-492-4. doi: http://doi.acm.org/10.1145/584792.584837.
53. F. Gasparetti and A. Micarelli. User profile generation based on a memory retrieval theory. In *Proc. 1st International Workshop on Web Personalization, Recommender Systems and Intelligent User Interfaces (WPRSIUI'05)*, pp. 59–68, (2005).

BIOGRAPHIES

Fabio Gasparetti is a Postdoctoral Fellow at the Artificial Intelligence Lab of the Department of Computer Science and Automation, University of "Roma Tre". His research interests include personalized search, user modeling and focused crawling.

Alessandro Micarelli is a Full Professor of Artificial Intelligence at the University of "Roma Tre", where he is in charge of the Artificial Intelligence Laboratory at the Department of Computer Science and Automation. His research interests include adaptive web-based systems, personalized search, user modeling, artificial intelligence in education.

Chapter 3

Unobtrusive User Modeling For Adaptive Hypermedia

Hlary J. Holz*, Katja Hofmann† and Catherine Reed‡

*Department of Mathematics and Computer Science
California State University, East Bay 25800 Carlos Bee Blvd.
Hayward, CA 94542, USA
*hilary.holz@csueastbay.edu
†katja.hofmann@gmail.com
‡catherine.reed@csueastbay.edu*

We propose a technique for user modeling in Adaptive Hypermedia (AH) that is unobtrusive at both the level of observable behavior and that of cognition. Unobtrusive user modeling is complementary to transparent user modeling. Unobtrusive user modeling induces user models appropriate for Educational AH (EAH) based on metaphors characterized by a shared locus of control over learning, such as small group learning. Transparent user modeling, on the other hand, induces user models appropriate for EAH based on metaphors characterized by an independent locus of control, such as social navigation.

We extend an existing decomposition model of adaptation in AH to incorporate summative evaluation of unobtrusive user modeling. Summative evaluation examines the effects or outcomes of a system, while formative evaluation is used to improve a system under development. We also introduce a separate, two-stage model for formative evaluation of unobtrusive user modeling. We then present results from a field study for the first of the two developmental stages and describe the current field study of the second developmental stage. Finally, we compare unobtrusive and transparent user modeling, and explain the role of each.

3.1. Introduction

3.1.1. *User modeling in adaptive hypermedia*

Adaptive Hypermedia systems (AH) adapt the presentation and navigation of web pages based on a variety of static and dynamic models.[1] Static mod-

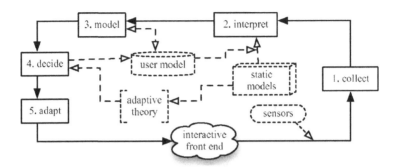

Fig. 3.1. A decomposition model of adaptation in AH, adapted from Paramythis and Weibelzahl.[19] This version is simplified in that it focuses on a single dynamic model, the user model.

els abstract information that does not change over the course of a single interactive session, such as models of domain knowledge, tasks, and the AH itself. Dynamic models capture information that changes more rapidly, such as user characteristics, behavior, interaction history and contextual information. User characteristics, behavior and interaction history comprise the *user model*.

Adaptation in AH can be decomposed into multiple stages (see Fig. 3.1).[2] The AH collects data via the interactive front end and/or non-interactive sensors. The data collected are interpreted with the aid of the static models and the user model. Based on the input data, any inferences drawn, and the existing user model, the AH constructs a model of the current state of the world. It may also update the user model. The AH decides what (if any) adaptation is needed given the current state of the world, based on its adaptation goals and objectives, or adaptive theory.[3] The AH adapts and the cycle repeats.

For example, a web-based Unix tutorial might track mouse and keyboard events. If a laboratory study was being performed, it might also collect eye-tracking data. Using pattern recognition techniques, the tutorial could infer the user's learning style from: the event data; the user's history; and hypermedia features of the current page, such as the presence of audio information. The tutorial would record its inference in the user model, as well as incorporate it into its model of the current state of the world. If the tutorial's adaptation goal was to help the user find materials best suited to her learning style, it might change the order or appearance of the links in a navigation area.

3.1.2. *Motivation: informal education and the user modeling effect*

Most existing Educational AH (EAH) are based on metaphors from formal education, such as lecture, textbook or tutoring. Our laboratory focuses on metaphors from informal education, especially small group learning modalities such as study groups and faculty office hours (see Sec. 3.3.1). The small group learning paradigm is characterized by: a shared locus of control over learning[a]; a lack of explicit assessment of the state of the user (for example, through quizzes or ratings); and an unpredictable learning environment. To adhere to the metaphor, we should: share control with the user; rely exclusively on implicit assessment; and support as broad a set of user-selected learning environments as feasible.

One approach to sharing control over learning with the user would be to use *transparent user modeling*.[4] Transparent user modeling exposes the adaptive process through the user interface, facilitating the user's development of mental models. However, research shows that awareness of the user model changes user behavior, a special case of the Hawthorne Effect.[5] For example, Claypool *et al.*, report that users asked to read and rate online items will spend time reading an item even if they do not find it interesting.[6]

In AH, one manifestation of the Hawthorne Effect is that awareness of user modeling tempts users to alter their behavior, consciously or subconsciously, a phenomenon we term the *User Modeling Effect* (UME). Although no current research directly evaluates the extent to which the UME skews data sets in the online environment, a general rule for the Hawthorne Effect is that the greater the awareness of being monitored and recorded, the stronger the effect.

3.1.3. *Our solution: unobtrusive user modeling*

An alternative way to share control is to develop an unobtrusive user model through iterative and participatory design. Rather than exposing a predetermined user model to the user, *unobtrusive user modeling* captures the natural structure of the user population. Unobtrusive user modeling lessens the impact of the UME by minimizing the user's continuing awareness of being monitored.

We have developed a technique for user modeling that is unobtrusive at both the data collection and cognitive levels. On the data collection

---
[a]When a learner expects to share control over the timing and shape of future learning with others, they are exhibiting a shared locus of control.

level, the user model is unobtrusive in that it collects data implicit in the learning process through current web technologies without altering normal user behavior or environment. On the cognitive level, the user model is unobtrusive in that it does not impose a predetermined structure on users, but rather employs the natural structure of the user population.

In this paper, we extend Paramythis and Weibelzahl's decomposition model to incorporate summative evaluation of unobtrusive user modeling. We also introduce a separate, two-stage developmental model for formative evaluation of unobtrusive user modeling. We then present results from a field study for the first of the two developmental stages and describe the current field study of the second developmental stage. Finally, we compare unobtrusive and transparent user modeling, and explain the role of each.

3.2. Approach

In their decomposition model, Paramythis and Weibelzahl distinguish between collecting data (stage 1) and making inferences from that data (stage 2) in order to facilitate formative and summative evaluation of adaptation in AH. The raw measurements collected in the first stage do not necessarily contain any semantic information. Semantic information is associated with the data at the second stage, when it is interpreted with the use of the static and dynamic models. Thus, recommended methods for evaluating the first stage concern the technical quality of the data acquired, as for example, reliability, precision and accuracy.[7] The proposed decomposition works fairly well when the raw measurements map directly to their semantically meaningful counterparts, called *features*.

For AH with inference grounded in statistical learning theory, however, it is important to distinguish among collecting data, extracting the subsymbolic structure of the data, inferring symbolic information from the subsymbolic structure, and modeling the state of the world. Statistical learning theory grounds the study of inference in a statistical framework, a set of "... assumptions of statistical nature about the underlying phenomena (in the way the data is generated)".[8] Figure 3.2 presents a revised model that explicitly includes each of these stages.

Distinguishing among collecting data, extracting subsymbolic structure, and inferring symbolic information are critical because the extraction process is dependent on how the extracted information will be used — the best features for one classifier are not necessarily the best features for a different classifier.[9] First, we need a set of measurements (in essence, minimally preprocessed features) that capture existing structure in the user population that the inference step can use. Next, we transform those measurements into a more compact, computationally tractable feature set. Thus we distinguish between *feature selection*, which is the process of selecting features that contain information relevant to the learning problem, and *feature extraction*, which is the process of computing new features or relations between features from existing features. In both cases, the goal is to reduce the dimensionality of the data with little or no loss of information.

3.2.1. *Classifier-independent feature selection*

Development of unobtrusive user modeling requires an additional, exploratory stage (see Fig. 3.3.) The goal of this exploratory stage, classifier-independent feature selection, is to construct an initial set of features that capture the natural structure of the user population present in the measurements, *independent of the assumptions of any particular inference algorithm.*[10]

We start by working with domain experts to construct a set of likely features, features we expect to be relevant under our adaptive theory. For example, if we wish to adapt to the user's learning style, likely features might include whether the user played video clips or sound files. Next,

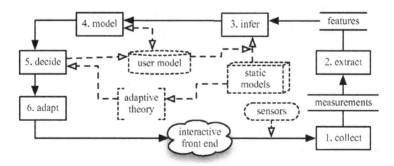

Fig. 3.2. A decomposition model of adaptation in AH that make inferences using statistical pattern recognition. (1) Collect measurements via the interactive front end and/or noninteractive sensors. (2) Extract features from those measurements that capture the natural structure of the user population. (3) Infer symbolic information about the user from the features with the aid of the static models and the user model. (4) Model the current state of the world and update the user model if necessary. (5) Decide what (if any) adaptation is needed. (6) Adapt the web page.

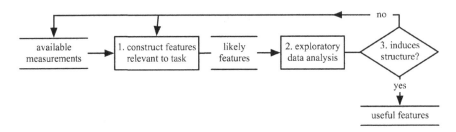

Fig. 3.3. Developmental Stage 1: Classifier-independent feature selection. (1) Design a set of features likely to extract information relevant to the adaptive task from the available measurements with the help of problem domain experts. (2) Because the structure of the data is poorly understood, analysis of the proposed features should use exploratory data analysis based on unsupervised learning. This step will require an initial training sample. (3) If the feature set does not induce sufficient structure, either redesign the features or seek out new measurements.

we perform an exploratory data analysis on those features using *unsupervised clustering*. In unsupervised clustering, the correct inferences for each sample in the data is unknown. The samples are analyzed on the basis of similarity between feature vectors. If the feature set does not induce sufficient structure, either the features may be redesigned or new measurements may be necessary.

Because the correct inferences are unknown, the feature set is evaluated purely on the quality of structure induced. How much and what kind of structural information are available varies by exploratory data analysis technique. Typically, exploratory data analysis techniques do not have quantitative measures of convergence or captured structure, but assist researchers in making a preliminary exploration of the feature space, typically through visualization or coarse approximations. What constitutes sufficient structure is a judgment call, although some guidance may be available from case studies using the particular exploratory data analysis technique.

Classifier-independent feature selection will not yield the best features for any one particular inference algorithm. Rather, it enables the developer to choose an inference algorithm on the basis of the natural structure of the user population, and to evaluate the measurements on a semantic level, independent of the subsequent steps.

3.2.2. *Inference design*

The result of the first developmental stage is a set of features that capture some of the natural structure of the user population, structure on which the AH should adapt. The goal of the next developmental stage, inference design (see Fig. 3.4), is to construct an inference algorithm that can exploit the features constructed in stage one to adapt the AH according to the given adaptive theory.

We start by identifying a set of candidate inference algorithms. Each inference algorithm considered will need a second round of feature analysis, comprised of classifier-specific feature extraction and possibly classifier-specific feature selection. Due to the confounding issues of classifier-independent versus classifier-specific features, statistical learning theory does not concern itself with finding the best inference algorithm, but with finding an inference algorithm that is "good enough".

Next, we use factorial design to set the parameters of the learning algorithm, including any feature extraction parameters. Factorial analysis will generally yield better results faster than heuristic and/or intuitive exploration of the design space. However, heuristics can be used to set the factor levels in the factorial design, as can the structural information from the previous stage. It is a good idea to use a new set of training data for factorial design, rather than the sample from the previous stage. If the inference algorithm is very closely related to the exploratory data analysis

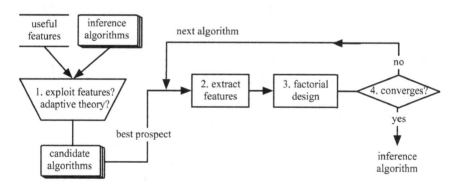

Fig. 3.4. Developmental Stage 2: Inference design. (1) Identify a set of candidate inference algorithms that exploit the features constructed in stage one (classifier-independent feature selection) and support the adaptive theory. Starting with the best prospect, extract a reduced set of classifier-specific optimal features from the existing features (2) and use factorial design to set any necessary parameters (3). If the feature extraction algorithm for the inference algorithm has parameters, steps two and three are not separable, and the extraction parameters should be included in the factorial design. (4) Select the design point for which the inference algorithm exhibits strongest convergence. If the inference algorithm does not converge for any design point, try the next candidate.

technique, using a new training sample is mandatory.

The inference algorithm is evaluated on the basis of convergence rather than performance because we want to evaluate the inference algorithm independent of the adaptive decision making. If the inference algorithm does not have a formal measurement of convergence, follow the accepted practice for that algorithm.

Once we have the inference algorithm, the classifier-independent stage is no longer necessary, and our developmental model reduces to the model in Fig. 3.2.

3.3. Field Study

We performed a field study of Developmental Stage One, classifier-independent feature selection, on an existing EAH. The results from the stage one field study are currently being used in a stage two field study.

3.3.1. *ACUT*

We tested our method in a field study on an existing EAH, the Adaptive Collaborative UNIX Tutorial (ACUT).[11] Originally, ACUT was conceived and implemented as an informal EAH intended to teach UNIX skills required for computer science (CS) studies. The current version is an open-source mod_perl webserver intended as an experimental platform for researching the small group learning metaphor in EAH.

ACUT was developed to address the problem of low retention of nontraditional students, for example, female and minority students, in CS.[12] Practical knowledge about the UNIX operating system is required for advanced undergraduate and graduate CS study. Acquiring competence in practical computing skills such as UNIX through formal education is problematic.[13] Traditional CS students acquire UNIX skills through informal education. Examples of informal education include discussions with friends or family members. Nontraditional CS students have less access to informal education in computing. To increase retention of nontraditional CS students, ACUT works to bridge both gaps — the difference in UNIX experience and the differential access to informal education in computing.

ACUT is a "meta-tutorial"; it helps students navigate the vast, unmanageable knowledge space of the web by filtering, organizing and annotating learning resources (see Fig. 3.5.) ACUT assumes the tasks of placing students' questions in context, and then recommending appropriate learning resources to students.

Beginning CS students often face the problems of not knowing what they do not know and where to start learning. One of the primary goals of ACUT is to improve peer-to-peer support through collaborative learning, both through explicit collaboration with other ACUT users and through implicit collaboration with other computer scientists in the web community. Students can add notes to each learning resource and also see other students' notes. ACUT is designed to adapt based on the subsymbolic cues that master teachers use in mentoring environments such as office hours or student research. Because ACUT is a research platform, however, the actual nature of adaptation in ACUT varies with time. Details concerning adaptive decision making (stages 4 and 5) are omitted from this discussion because our current focus is on the development of the user model.

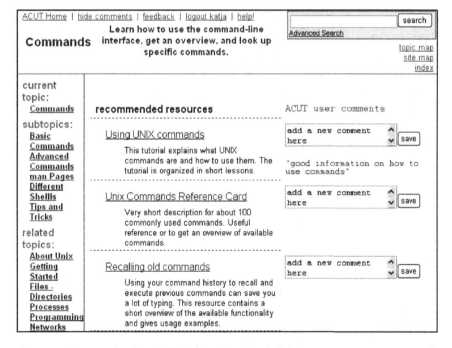

Fig. 3.5. A screenshot from ACUT (the Adaptive Collaborative UNIX meta-Tutorial).

3.3.2. Measurements

In prior research, we developed a technique to unobtrusively collect a rich set of data on subsymbolic user behavior in AH.[14] We used our technique to record user behavior and page characteristics as follows:

- User behavior
  - (a) page requests
  - (b) mouseovers
  - (c) clicks
  - (d) pauses (periods of inactivity greater than two seconds)
  - (e) scrolling
  - (f) resize events
  - (g) tool use (modify, edit or create comments).
- Page characteristics
  - (a) tag map[b]
  - (b) size (bytes sent)
  - (c) URI
  - (d) number of topics
  - (e) number of related topics
  - (f) number of subtopics
  - (g) number of resources
  - (h) number of own comments
  - (i) number of total comments.

3.3.3. Feature design

From these measurements, we constructed an initial feature set based on reports in the literature, ACUT's existing design documentation, and our intuition. For example, *time spent reading*, mouseovers, clicks, and scrolling have shown some promise in capturing subsymbolic user behavior.[15–17] We constructed an initial set of self-normalizing features incorporating these measurements. These features are all fairly generic, applicable to a broad set of EAH.

Initial feature set

- event share–mouseover: mouseovers as % of total events

---
[b]A tag map is a string representation of all the names of the HTML elements in a page and the number of times each occurs.

- event share–resize
- event share–scroll
- event share–click
- time spent reading: % of possible time spent reading (max of 30 min)
- pause frequency[c]
- activity time: time spent reading — total time paused.

3.3.4. *Data collection*

Users can register online for ACUT.[11] The registration process involves three steps: facilitating informed consent for human subjects research; completing a questionnaire; and creating a unique username and password. The contents of the questionnaire vary over time according to the needs of current ACUT-supported research projects.

The results in the following sections are for data generated by 22 users in 35 sessions between April 18th and May 7th, 2006. The users visited 698 pages within ACUT, producing a total of 3278 trace events. Approximately half of the users were age 20 to 29, the rest were older. Just under 20% were female. Approximately half were undergraduates. Two-thirds were CS majors, with the rest drawn from other hard sciences. Most reported using computers several hours each day. None rated themselves as beginners at computing, but rather as intermediate or expert. In contrast, well over half rated their UNIX expertise as either beginner or advanced beginner. Only a few included friends or family in their primary or secondary sources for learning UNIX. Most users were, in fact, nontraditional in one sense or another, as all but two were from our institution, which specializes in nontraditional students.

3.3.5. *Self-organizing maps*

For data about which we have very little knowledge, we need to use exploratory data analysis.[18] The goals of exploratory data analysis are (1) to gain a better understanding of the structure of the data and (2) to identify previously unknown patterns and relationships within the data.

We use self-organizing maps (SOMs) for exploratory data analysis, as SOMs can be used to visualize fine structures and clustering tendencies underlying high-dimensional data.[18,19] SOMs consist of a grid of map units (or neurons), each of which is associated with a weight vector of the same

---

[c]Inactivity of two seconds or longer was recorded as a pause, therefore the maximum number of pauses is (time spent reading/2).

dimensionality as the input data. The weight vectors are randomly initialized. The SOM is trained in several phases, so that the weight vectors of the map units approximate the input data in an ordered way.

Map training minimizes the average quantization error, defined as the average Euclidean distance of each data sample from the map unit closest to that data sample. Training several SOMs on a data set with identical parameters will yield different maps, depending on map initialization. The quantization error is used to select the best trained map. Graphing the average quantization error is used to estimate convergence (see Fig. 3.6), because SOMs lack a formal measure of convergence.

The trained SOM is then transformed in a variety of ways to visualize different aspects of the input data. Several implementations for training SOMs and generating map visualizations are available. We used the program package SOM_PAK.[20]

To visualize the structure of our data, we use Sammon mapping (see Fig. 3.7).[21] Sammon mapping projects points in high-dimensional data to lower dimensionality, usually a two-dimensional plane. Sammon maps preserve topology in that the projection of neighboring points approximates the distances between the points in the original dimensionality. Sammon maps are complementary to SOMs, as the former visualize distances between data points while the latter preserve neighborhood relationships and visualize these in an ordered way. Different algorithms for generating Sammon

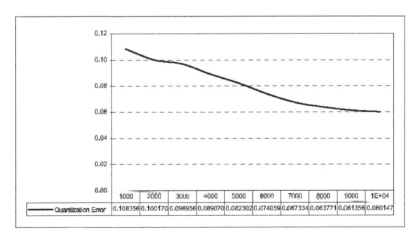

Fig. 3.6. Convergence of the quantization error for the SOM trained on the initial feature set. The table below the graph shows the quantization error for map "snapshots" every 1000 training steps.

maps exist. We use SOM_PAK's implementation of the iterative algorithm described in Sammon.[21]

Figure 3.7 shows the Sammon maps for our initial feature set. In combination with the quantization error graph, the Sammon maps show that we have collected sufficient training data to proceed, and that the map parameters chosen are appropriate for our data.

To visualize the clusters in the trained SOM, we used a U-matrix representation.[22] To build a U-matrix representation, also called a cluster tendency map, we find the closest map unit for each sample in the input data. Each map unit is then colored (gray level) depending on how many input samples are closest to that unit. The fewer samples a map unit "wins", the darker the color of the unit in the U-matrix. The U-matrix is then blurred. Areas of the map corresponding to high density of input samples will be light or white. These areas constitute likely clusters within

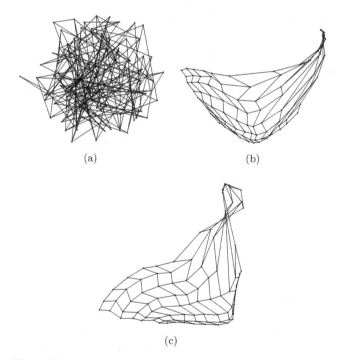

Fig. 3.7. We use Sammon maps to inspect the structure of the SOM before, during and after map training. (a) The distances between weight vectors are randomly initialized. (b) After a first training run, the neighboring weight vectors are ordered. (c) Subsequently, the SOM is trained to reflect the finer structure of the input data. Points projected in close proximity indicate a clustering tendency.

the data. Dark areas in this representation are of low density and represent likely cluster boundaries.

Figure 3.8(a) shows the cluster tendency map for our initial feature set. The data exhibit little clustering tendency. The map has one cluster, labeled '1', a very pronounced cluster in the lower-left hand corner.

To understand how each input feature contributes to clustering in the trained SOM, we use component "planes".[19] A separate visualization is generated for each of the input features. Map units with high values for a particular feature are represented as light values of gray. Map units with low values are represented as dark. The resulting representations show the distribution of each feature on the trained SOM, and are used to interpret the influence of each feature on the global and local ordering of the SOM.

To interpret the component planes, we look for structures within the component planes that best explain local clustering tendencies and cluster boundaries within the U-matrix representation.[23] In addition, the global ordering of a feature can indicate the importance of that feature for global ordering of the SOM.

Figures 3.8(b)-8(h) show the component planes for our initial feature set. Focusing on the three features that contribute to the cluster structure, we see that cluster 1 is characterized by high activity time (c), but low values for event share–click (d) and event share–mouseover (b). We can also see that the global structure in the activity time component plane is reflected in the global structure of the cluster tendency map. Time spent reading (f), event share–resize (g) and event share–scroll (e) also show structure in the component planes, but do not seem to have much influence on either local or global structure of the SOM.

3.3.6. *Revising the features*

We combined the most promising features in the initial set with new features as described below to generate a revised feature set. To revise the features, we worked with a domain expert, a cognitive scientist working in internet communication. Thus the results of the first exploratory data analysis phase established the basis for the participatory design process.

We retained activity time, as it contributed significantly to both global and local structure in the first SOM.

Time spent reading and scrolling both induced some structure in the initial data. Taken separately, neither had much influence on the cluster structure of the map. Working with our domain expert, we constructed

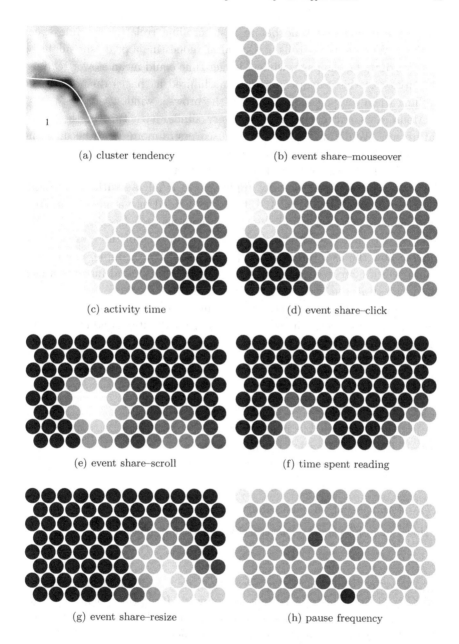

(a) cluster tendency (b) event share–mouseover

(c) activity time (d) event share–click

(e) event share–scroll (f) time spent reading

(g) event share–resize (h) pause frequency

Fig. 3.8. The cluster tendency and component plane maps for the initial feature set. (a) The initial feature set does not induce very much structure (compared to Fig. 3.10). Of the structure induced, event share–mouseover and event share–click are the strongest contributors, while pause frequency is the weakest contributor.

a new feature, named time before first scrolling (relative to time spent reading). When a user scrolls quickly after requesting a page, she might be quickly skimming the page, while a longer time could mean slower reading. The utility of this feature is somewhat limited as pages do not require scrolling when the page contents fit in the browser window.

Mouseovers and clicks exhibited very similar structure in the initial feature set. We decided to explore mouseovers in more detail because the data contained far more mouseovers.

Combining mouseovers with hypermedia information, we constructed new, richer features still broadly applicable to EAH. As with many EAH, the structure of the pages in ACUT can be divided into a navigation area, an administrative area (topic bar) and a content area (see Fig. 3.5). We hypothesized that the amount of activity within each of those areas could indicate how the user interacts with web pages. For example, moving the mouse over a large percentage of the navigation area could indicate a user who looks at all the options carefully before choosing one.[d]

Combining the locus of activity with time before first scrolling anchored the revised feature set. The revised feature set exhibited good convergence and far more structure than the initial feature set (see Fig. 3.9).

Revised feature set

- Where on the page is the user most active?
  (a) % links on topic bar moused over
  (b) % links on navigation bar moused over
  (c) % links in content moused over.
- How does user interact with text?
  (a) % time spent on page before first scrolling.
- How much does the user interact with each web page?
  (a) % activity time.

Analyzing the U-matrix representation, we can distinguish seven clusters (Fig. 3.10). Cluster 1 is very pronounced. It captures behaviors with high activity time and time before first scroll but few mouseovers of any kind. Cluster 2 is a large cluster that is not clearly distinguished by boundaries in the component planes. Feature values are: high time before first scroll and activity time; low navigation link mouseover; and mixed topic

---
[d]Mousing over a navigation link reveals a description of the target page.

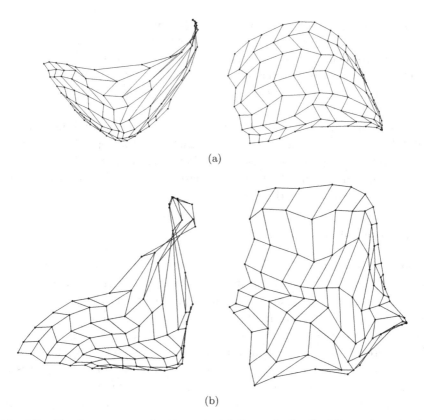

Fig. 3.9. The Sammon maps for the revised feature set. On the left are the ordered and fully trained Sammon maps for the initial feature set. On the right are the same maps for the revised feature set. (a) An improvement over the initial feature set can already be seen after the first training run. (b) After fine training, the SOM for the revised feature set shows a dramatic increase in structure.

link and content link mouseover. Adding features might reveal additional structure in this cluster. Cluster 3, less pronounced, is characterized by mid-range navigation link mouseover, mid to high content link and topic link mouseover, high time before first scroll and low activity time. As with cluster 2, additional features might reveal additional structure. Cluster 4 is a pronounced cluster, characterized by high time before first scroll, low navigation link mouseover, content link mouseover, low to mid topic link mouseover and low activity time. Cluster 5 is also fairly well pronounced. Like cluster 4, it captures behaviors with high time before first scroll and low navigation link and content link mouseover. However, these behaviors have low topic link mouseover and mid-range activity time. Cluster 6 is another very pronounced cluster with high navigation link mouseover as the identifying characteristic. Cluster 7, which has low time before first scroll as the identifying characteristic, is not very pronounced. In the current field study, we no longer collect time before first scroll unless the user scrolls on a page. We expect this change to improve the strength of time before first scrolling.

Incorporating hypermedia information into mouseovers produced three useful features: % navigation links moused over; % topic links moused over; and % content links moused over. We hypothesize that including hypermedia information will also differentiate mouseovers from clicks, which hypothesis we are currently testing.

3.4. Discussion

The process described allows us to visualize the structure underlying our data set. We constructed a feature set for a live EAH based on cluster visualization and expert analysis. Analyzing this feature set, we identified previously unknown patterns of user behavior. Thus, through iterative and participatory design, we have identified a small set of behavior patterns that represent the ways in which users typically interact with ACUT.

These revealed behavior patterns suggest questions for further investigation. These questions include: What symbolic information can we attach to these behavior patterns, if any? When do they occur? Which students exhibit which behavior patterns on which pages?

One way to analyze the semantic content of the SOMs is to attach symbolic labels to the map units. For a set of labeled input samples, we find the map unit closest to the input sample. The map unit is then assigned the label of the input sample. Some map units might not be associated with a label, while other map units might be associated with several labels. In the latter case, we only display the label of the closest sample. Labeled SOM units can be used with either cluster tendency or component plane maps.

The labeled maps in Fig. 3.11 are typical of our data set. None of the user characteristics currently collected in the registration process explains the global or local structure of the trained SOM.[e] Labeled cluster tendency maps can also be used to estimate the magnitude of the transformation required in order to use the features to infer the symbol. While the revised feature set captures significant structure, strong feature extraction will be

---
[e]For a contrasting example, see Ref. 23.

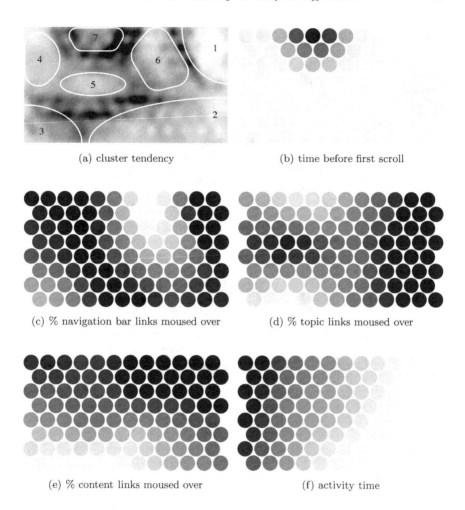

Fig. 3.10. The cluster tendency and component plane maps for the revised feature set. The revised features induce far more structure on the user population (compared to Fig. 3.8). All the features contribute to the induced structure. Time before first scroll, % of navigation bar links moused over, and % of topic links moused over can be used as currently constructed. Activity time and % content links moused over merit (and require) further exploration.

required to infer any of the desired symbolic values.

We are currently conducting a field study of Developmental Stage Two, inference design, on ACUT. Knowing that the revised feature set exhibits strong global and local structure, we can design an inference engine to

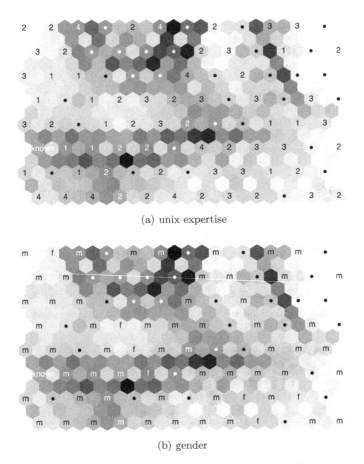

Fig. 3.11. Cluster tendency maps labeled by symbolic feature. If a symbol's values cluster in the same way as the gray-scale values cluster, minimal feature extraction will be needed to infer that symbol from the feature set. (a) Unix expertise at time of registration, ranging from beginner (displayed as 1) through expert (displayed as 4). (b) Gender. Significant feature extraction will be required to infer either of the desired symbols.

exploit that structure. The inference engine will infer the user's learning style from the revised feature set. Once we have a successful inference engine, we plan to explore using the inferred learning styles to recommend resources and to facilitate study networks.

Both transparent and unobtrusive user modeling are important in that they incorporate the task-artifact cycle[f] into the design methodology for an

---

[f]Introducing a computational artifact changes the way users perform a task, a phe-

EAH. Perhaps even more important is that both methods can be used to contribute to "... understanding ... the distinctive dynamics of working with computational artifacts, and ... their artful integration with the rest of the social and material world".[25]

The dynamics of working with computational artifacts depend, in part, on which method is employed. Neither transparent nor unobtrusive user modeling is intrinsically preferable, rather, they induce different kinds of user models. Transparent user modeling induces a user model that is appropriate for EAH based on metaphors with independent locus of control over learning, such as indirect social navigation. Unobtrusive user modeling induces a user model that is appropriate for EAH based on metaphors with shared locus of control, such as small group learning. The two can also be combined, for example, by transparently modeling goals and unobtrusively modeling behavior.

The distinction between transparent and unobtrusive user modeling is fundamental to our research. In transparent user modeling, the user is able to "see" that actions are being captured. The data is affected by user awareness, triggering the User Modeling Effect, although the extent of the effect has not been measured. In unobtrusive user modeling, conscious awareness of being monitored fades over time, as the user "sees" neither the process of monitoring nor the technical details of the relationship between monitoring and adaptation in the EAH. Thus the structure in the user population that forms the basis for unobtrusive user modeling remains relatively unchanged by the User Modeling Effect. This relatively unchanged structure will enable the construction of more robust EAH, scaffolding the learning needs of all students regardless of their expertise.

Acknowledgments

We would like to thank our domain expert, Dr. John Lovell. Katja Hofmann was supported by a Fullbright Scholarship and a CSU, East Bay Associated Students fellowship. We also thank Robert Sajan and Ifeyinwa Okoye for proof-reading.

References

1. P. Brusilovsky. Methods and techniques of adaptive hypermedia. In eds. A. K. P. Brusilovsky and J. Vassileva, *Adaptive Hypertext and Hypermedia*, pp. 1–

---

nomenon known as the task-artifact cycle.[24]

34. Kluwer, (1998).
2. A. Paramythis and S. Weibelzahl. A decomposition model for the layered evaluation of interactive adaptive systems. In *Proc. User Modeling 2005, 10th Int. Conf.*, pp. 438–442, Edinburgh, Scotland (July, 2005).
3. P. Totterdell and P. Rautenbach. Adaptation as a problem of design. In eds. P. T. D. Browne and M. Norman, *Adaptive User Interfaces*, pp. 61–84. Academic Press, (1990).
4. K. Höök, Steps to take before intelligent user interfaces become real, *Interact. Comput.* **12**(4), 409–426, (2000).
5. H. Parsons, What happened at hawthorne?, *Science.* **183**, 922–932, (1994).
6. M. W. M. Claypool, P. Le and D. Brown. Implicit interest indicators. In *Proc. 2001 Int. Conf. Intelligent User Interfaces*, pp. 33–40, Santa Fe, NM, USA. (January, 2001).
7. C. Gena, Methods and techniques for the evaluation of user-adaptive systems, *Know. Engin. Rev.* **20**, 1–37, (2005).
8. S. B. O. Bousquet and G. Lugosi. Introduction to statistical learning theory. In eds. U. v. L. O. Bousquet and G. Ratsch, *Advanced Lectures on Machine Learning*, pp. 169–207. Springer, NY, (2003).
9. M. Ben-Bassat. Pattern recognition and reduction of dimensionality. In eds. P. Krishnaiah and L. Kanal, *Handbook of Statistics*, pp. 773–791. Elsevier, Amsterdam, (1982).
10. H. J. Holz. *Classifier-independent feature analysis*. PhD thesis, George Washington University, Washington, DC (May, 1999).
11. H. J. Holz. The adaptive collaborative unix tutorial. `http://acut.csueastbay.edu/`.
12. R. Farzan. Adaptive collaborative online unix tutorial for computer science students. M.s. thesis, CSU, East Bay, (June 2003).
13. C. Tully, Informal education by computer — ways to computer knowledge, *Comput. Edu.* **27**(12), 31–43, (1996).
14. C. R. K. Hofmann and H. J. Holz. Unobtrusive data collection for web-based social navigation. In *Proc. Workshop on Social Navigation and Community-Based Adaptive Technologies*, Dublin, Ireland (June, 2006).
15. S. Bidel. Statistical machine learning for tracking hypermedia user behavior. In *Proc. UM '03 Workshop on Machine Learning, Information Retrieval and User Modeling*, pp. 56–65, Johnstown, PA (June, 2003).
16. J. Goecks and J. W. Shavlik. Learning users' interests by unobtrusively observing their normal behavior. In *Proc. 2000 Int. Conf. Intelligent User Interfaces*, pp. 129–132, New Orleans, LA (January, 2000).
17. Y. Hijikata. Implicit user profiling for on demand relevance feedback. In *Proc. 2004 Int. Conf. Intelligent User Interfaces*, pp. 198–205, Funchal, Portugal (January, 2004).
18. J. N. S. Kaski and T. Kohonen. Methods for exploratory cluster analysis. In *Proc. Int. Conf. Advances in Infrastructure for Electronic Business, Science, and Education on the Internet.*, (2000).
19. T. Kohonen, *Self-Organizing Maps*. (Springer-Verlag, NY, 2001).
20. T. K. *et al.* Som pak: the self-organizing map program package, (1996).

21. J. Sammon, A nonlinear mapping for data structure analysis, *IEEE Trans. Comput.* **C-18**(5), 401–409, (1969).
22. A. Ultsch. Self-organizing neural networks for visualization and classification. In eds. B. L. O. Opitz and R. Klar, *Inform. Classification*, pp. 307–313. Springer, (1993).
23. J. N. S. Kaski and T. Kohonen. Methods for interpreting a self-organized map in data analysis. In *Proc. 6th European Symp. Artificial Neural Networks*, pp. 185–190, Bruges, Brussels (April, 1998).
24. W. K. J. Carroll and M. Rosson. The task-artifact cycle. In eds. J. M. Carroll and J. Long, *Designing Interaction: Psychology at the Human-Computer Interface*, pp. 74–102. Cambridge University Press, (1991).
25. L. A. Suchman. From interactions to integrations. In *Proc. IFIP TC13 Int. Conf. Human-Computer Interaction*, p. 3, Sydney, Australia (July, 1997).

BIOGRAPHIES

Hilary Jean Holz received the B.S. degree in mathematics and computer science from Dickinson College, Carlisle, PA, in 1984, and the M.S. (1989) and Ph.D. (1999) degrees in computer science from the George Washington University, Washington, D.C. She is an Assistant Professor at Cal State, East Bay.

Catherine Reed received a B.A. in German from Pomona College in 1966, an M.A. in German linguistics from San Jose State University in 1972, and a Ph.D. in educational psychology from the University of Virginia in 2001. She is an Assistant Professor at Cal State, East Bay.

Katja Hofmann received the degree DiplomInformatiker from the University of Applied Sciences (HTW), Dresden, Germany, in 2003, and the M.S. degree in computer science from Cal State, East Bay, in 2006. She is currently working as a scientific programmer at the University of Amsterdam.

Chapter 4

User Modelling Sharing for Adaptive e-Learning and Intelligent Help

Katerina Kabassi,* Maria Virvou[†] and George A. Tsihrintzis[†]

*Department of Ecology and the Environment, Technological Educational Institute of the Ionian Islands, Zakynthos 291 00, Greece
e-mail: kkabassi@teiion.gr
[†]Department of Informatics, University of Piraeus, Piraeus 185 34, Greece
e-mail: { mvirvou, geoatsi } @unipi.gr

This paper presents a domain independent method for user modelling sharing between applications. More specifically, a user modelling server, called UM-Server, has been developed. UM-Server embodies a multi-criteria decision making theory on the server side and provides services to any application that requests for it. To demonstrate the re-usablility of UM-Server, we have used a common user model in three different applications of various domains, an e-learning system for health issues and two intelligent user interfaces, one for file manipulation and one for e-mailing. Despite the differences that exists between these domains, it is shown that the generalised method proposed can be successfully used by various applications.

4.1. Introduction

Most people interacting with a computer program encounter problems that are related to the system usage. This problem may be addressed if the system is adapted to goals, needs and interest of each individual user. However, this adaptation is possible only if some characteristics of the user interacting with the software are taken into account. These characteristics are used to build some kind of a user model.[1,2] User models are very useful because they provide an insight to software applications about the real intentions of users, their preferences and habits, their level of knowledge and beliefs and possible misconceptions or unintended actions. This information is used by user modelling mechanism to improve the system's understanding of the user interacting with it at the particular moment.

However, the main problem with this approach is that the system needs quite a long time before it gathers adequate information about the user's goals, errors, misconceptions or interests. This problem has been addressed in the past by using stereotypes. Indeed, stereotypes constitute a powerful mechanism for building user models[3] and have been widely used in advisory software[1,2,4-6] and intelligent tutoring systems.[7] More specifically, stereotypes are used in user modeling in order to provide default assumptions about individual users belonging to the same category according to a generic classification of users that has previously taken place.[1,2] These inferences must, however, be treated as defaults, which can be overridden by specific observations which also requires the user interacting with the system for quite a long time.

Such problems may be overcome if several applications use a common user model. As a result, every time the user uses a new application, this application will have the ability to adapt to his/her characteristics, as these would be available from the common user model. These common user models may be provided by user model server. Indeed, a user model server could enable the reuse of the user model across applications,[8] and, as a result, improve the user modelling procedure. For a comprehensive review and an analysis of commercial user modelling servers the reader can refer to Fink and Kobsa.[9]

In view of the above, we have developed a User Modelling server. The developed server is called UM-Server and its main characteristic is that is uses a decision making theory for user modelling. UM-Server has been tested for providing information for intelligent help and adaptive e-learning. For this purpose, UM-Server has been used by three two different applications, an e-Learning system (INTATU) for a disease called Atheromatosis, an emailing system (MI-Mailer) and a file manipulation program (MBIFM), which share the same user model. The domain of an e-learning system about Atheromatosis differentiates in many ways from the domain of intelligent user interfaces that provide intelligent help during file or e-mail manipulation.

All these systems constantly reason about every user action and provide spontaneous advice in case of an error. This kind of reasoning is performed by the user modelling component that communicates with the UM-Server. In case one of these systems judges that the user may have made a mistake with respect to his/her hypothesised intentions, it suggests an alternative action that the user may have really meant to issue rather than the one issued. The selection of the best alternative is a multi-attribute decision problem. Therefore, both systems use Simple Additive Weighting (SAW)[10,11] for selecting the best alternative. The values of the attributes needed for the application of SAW are provided by the user model maintained in the UM-Server.

Multi-criteria decision making theories are not easily adapted in Intelligent User Interfaces (IUIs). As a result many development steps are required for their effective application. These steps involve many experimental studies and complicate the knowledge-based software life-cycle.[12] In this sense, the development of a Web Service that incorporates the reasoning of a decision making theory would be very beneficial for developers of adaptive user interfaces. We have tested the reusability of UM-Server in two different applications, a file manager and an e-mail client.

UM-Server uses Web Services for information sharing. Web Services introduced a new model on the Web, in which information exchange is conducted more conveniently, reliably and easily. More specifically, Web Services are self-contained, modular applications via the Web that provide a set of functionalities to anyone that requests them. One of the advantages of Web Services is that they allow a higher degree of reusability of components and knowledge sharing. This is very important for software that is quite complex to develop from scratch. This is certainly the case for user modelling components. However, the technology of Web Services by itself cannot ensure reusability if the application has not been designed in an appropriate way that makes optimum use of the potential offered by the technology.

The main body of this chapter is organized as follows: In Section 2 we describe the three systems that share the common user model. More specifically, it is presented an e-Learning system for health issues, an Intelligent User Interface for an e-mailing system and an Intelligent User Interface for file manipulation. In Section 3 we present and discuss the attributes that are commonly used by all applications for evaluating alternative actions. In Section 4, we give examples of a user's interaction with all three applications and how the common user model is used. In Sections 5 and 6, we describe the UM-Server's functionality and interaction with the client applications. Finally, in Section 7 we give the conclusions drawn from this work.

4.2. Description of Systems of Different Domains Sharing a Common User Model

4.2.1. *System for e-Learning in Atheromatosis*

E-learning and e-health can help users take more control of their well-being and improve their lives by accessing health information. However, this information is sometimes inaccessible for most people, as they do not have the background knowledge to understand the medical terminology used. A solution to this problem may be achieved by providing each user with personalized information that is tailored to his/her knowledge and interests.

In view of the above, we have developed an e-learning system for a medical domain that has the ability to adapt its interaction to each user dynamically. The system is about Atheromatosis, which is a topic that is of interest to many categories of people. Atheromatosis of the aortic arch has been recognized as an important source of embolism. System embolism is a frequent cause of stroke. The severity of Atheromatosis is granted by the fact that aortic atheromas are found in about one quarter of patients presenting embolic events.[13] Information about Atheromatosis is considered crucial because the diagnosis of this particular disease is mostly established after an embolic event has already occurred.

The e-learning system developed is called INTATU (INTelligent Atheromatosis TUtor). The system addresses a variety of users, such as patients, patients' relatives, doctors, medical students, etc. The main goal of IN-TATU is to adapt dynamically its interaction to each user. The main problem that users of such an application face is that most of them are not familiar with interacting with a computer and as a result their learning process is slowed down. A solution to this problem may be given if the system provided them support during their learning process.

4.2.2. *Systems for Intelligent Help in file manipulation and e-mailing*

Recently there have been several approaches for intelligent help which all aim at improving the quality of help to the user.[14] Very often, help is given after an explicit user's request like in UC,[5,6] which is an intelligent help system for Unix users. However, one important problem that has been revealed by empirical studies (e.g.[15]) is that users do not always realize that they have made an error immediately after they have made it. Therefore, they may not know that they need help. This problem can be addressed by active systems that intervene when they judge that there is a problematic situation without the user having initiated this interaction.

In view of the above, two different Graphical User Interfaces, MBIFM and MI-Mailer, have been developed: MBIFM is a file manipulation system and MI-Mailer is an e-mail client. Both systems aim at alleviating the frustration of users caused by their own errors and increasing their productivity. This is achieved by constantly reasoning about every user action and, in case of an error, providing spontaneous advice.

4.2.3. *Error Diagnosis in three systems of different domains*

All three systems described above aim at supporting users during their interaction with the computer and operate in a similar way. In particular, the common aims and operation of each of the three systems (MBIFM, MI-Mailer or INTATU) can be summarised as follows:

The system monitors every user's actions and reasons about them. In case the system diagnoses a problematic situation, it provides spontaneous advice. When the system generates advice, it suggests to the user a command, other than the one issued, which was problematic. In this respect, the system tries to find out what the error of the user has been and what the user's real intention was. Therefore, an important aim of the system's reasoning is error diagnosis. The reasoning of the system is largely performed by the shared user modelling component, which tries to model the user in terms of his/her possible intentions and possible mistakes. More specifically, the system evaluates the actions of every single user in terms of their relevance to his/her hypothesized goals. In case an action contradicts the systems' hypotheses about the user's goals or if it is wrong with respect to the user interface formalities, the system tries to generate a hypothesis about which action the user really intended to issue. During this process many alternative actions may be generated. The alternative actions that the system generates are similar to the user's initial action and compatible to his/her initial action. In order to evaluate the possible alternatives, rank them and select the one that is more likely to have been intended by the user, the system uses an adaptation of a multi-criteria decision making theory, which is shared by all systems described above (MI-Mailer, MBIFM and INTATU). The particular theory that has been adapted and incorporated in the user modelling mechanism is SAW, which is briefly described in the Appendix at the end of this document.

The actual adaptation and application of a multi-criteria decision making theory is a demanding procedure that requires many development steps.[12,16,17] Therefore, we have adapted a Web Service User model that maintains user models and incorporates a decision making theory. These user models are used by all systems described above (MI-Mailer, MBIFM and INTATU). In this way, information collected during the interaction of the user with MI-Mailer or MBIFM can be also used for the same users' interaction with INTATU and vice versa.

In case a user is new to both systems and, thus, there is not any information stored about him/her, the system uses stereotypes in order to initialise the user model for the new user. User stereotypes are used in order to provide default assumptions about users until systems acquired sufficient information about each individual user. In UM-Server, the default assumptions of the stereotypes are given in the form of the values of attributes that are essential for the application of SAW. Indeed as[1,2] points out, a stereotype represents information that enables the system to make a large number of plausible inferences on the basis of a substantially smaller number of observations; these inferences must, however, be treated as defaults, which can be overridden by specific observations. Therefore, MI-Mailer, MBIFM and INTATU constantly observe their users, collect information about them and send it to UM-Server that updates the individual user models.

4.3. Common attributes for evaluating alternative actions

In order to locate the attributes that human experts take into account while providing individualised advice, we conducted an empirical study.[12] During this process 16 human experts were reviewed about the attributes that they take into account when providing advice in computer applications. The attributes that were selected for the application of the multi-attribute analysis method were proposed among other attributes by the majority of human experts that participated in the experiment and are the following:

(1) Frequency of an error (f): The value of this attribute shows how often a user makes a particular error. Some users tend to entangle similar objects, other users entangle neighbouring objects in the graphical representation of the file store and others mix up commands. As the frequency of an error increases, the possibility that the user has repeated this kind of error increases, as well.
(2) Percentage of the wrong executions of a command in the number of total

executions of the particular command (e): The higher the number of wrong executions of a command, the more likely for the user to have failed in the execution of the command once again.

(3) Degree of similarity of an alternative action with the actual action issued by the user (s): Similar commands or objects of the file store are likely to have been confused by the user. Therefore, the similarity of the object and the command selected with the object and the command proposed by the system is rather important in order to locate the user's real intention.

(4) Degree of difficulty of a command (d): It has been observed that some commands are not easily comprehensible by the user. Therefore, the higher the degree of difficulty of a command or topic, the more likely for the user to have made a mistake in this command or want to revise the topic.

(5) Degree of relevance to the user's goals (g): An alternative action may be proposed to a user if it confirms the user's goals and plans or if it does not influence them. The actions that complete or continue an already declared and pending plan have higher degree of relevance to the user's goals than other actions.

The attributes f and e acquire their values directly from the information that is stored in the user model that is maintained centrally on the server. The user model contains information about the user's goals and plans as well as his/her errors and misconceptions. This information can be used by all systems described above (MBIFM, MI-Mailer and INTATU). As a result, if a user has interacted with MBIFM and is prone to accidental slips, this kind of information could be used when the user interacts with MI-Mailer or INTATU, as well. The value of the attribute g is acquired by the short term user model that is maintained locally on the user's computer. The value of the attribute d is predefined and static and is connected to each command. Finally, the value of the attribute s is dynamically calculated and is connected not just to the command issued in the graphical user interface of the client application but to the graphical representation of the user's file-store state or the user's mailbox as well. An example operation of each system is described in the next Section.

4.4. Example of a user interacting with three different systems

In this Section, we give examples of how a common user model stored in UM-Server is used in a user's interaction with MI-Mailer, MBIFM and INTATU for the same user. In the specific example, UM-Server indicates that the particular user is an expert in the usage of the computer, but rather careless:

In the case of INTATU, the user is presented several theory topics and the user selects to access the theory topic "Cardiography". However, the system after error diagnosis suspects that this theory topic is probably not the one that s/he really wanted to access as the particular user is novice with respect to his/her knowledge in Atheromatosis and does not have the prerequisite knowledge for this subject. Therefore, INTATU generates the following two alternative actions:

T1. Cardiology

T2. Computed Tomography

Both topics are neighbouring in the list presented to the users and their names are quite similar. For every alternative action that has been generated, the system uses the information from the user model that is stored on the Server in order to calculate the values of each criterion. This information is then sent back to the Web Service, which is responsible for applying SAW and selecting the best alternative. In the particular case the best alternative was the first one. Table 4.1 presents the values of the criteria for every one of the three alternative actions. Furthermore, it updates the information stored in the user model maintained on the UM-Server.

Table 4.1. The values of the criteria based on the information stored in UM-Server user model.

	f	e	s	d	g
T1	0.7	0.3	0.9	0.1	0.9
T2	0.5	0.3	0.6	0.1	0.3

In the case of MI-Mailer, the example initial state of the user's electronic mailbox is illustrated in Figure 4.1. The user has just received a possibly infected message and tries to delete the message in a way that she would not activate the virus. Since the user is an expert, she knows that in some of the most popular e-mailing systems users can only avoid a virus sent by e-mail by not opening or even selecting the message. Indeed, in such clients, selecting a message to display it in a preview frame triggers the execution of possible scripts that it may contain. Furthermore, in most e-mailing systems the only way of deleting an e-mail message is by selecting it and then executing the command delete. Therefore, the user in the particular case, like the majority of expert users, finds different ways of deleting an infected message without selecting it.

Fig. 4.1. The user's electronic mailbox

The user's final goal is to delete the infected message, which is located in the folder 'Inbox/Course2', without 'clicking' on it. In order to achieve this, she intents to move the rest of the messages, which are rather important for her, into another folder and then, delete the folder which contains only the infected message. Therefore, she moves all the messages stored in the folder 'Inbox/Course1' except for the infected message to the newly created folder.

However, she accidentally tries to delete the folder 'Inbox/Course1'. This action would be disastrous because the user would have deleted all the messages that she intended to keep. Therefore, the system finds the particular action 'suspect' and generates alternative actions. The alternative actions that are generated by the system are the following:

J1. delete(Inbox\ Conferences\)
J2. delete(Inbox\Course2\)
J3. delete(Inbox\Courses\)

The alternative actions result from the assumption that the user probably wanted to delete a different folder and, therefore, the system proposes the folder 'Inbox/Course2' that has the infected message and the folders 'Inbox/Courses' and 'Inbox/Conferences' that are empty. Similarly to IN-TATU, the system uses the information from the user model that is stored on the Server in order to calculate the values of each criterion for each one of the alternative actions presented above. This information is then sent back to the Web Service, which is responsible for applying SAW and selecting the best alternative. In the particular case the best alternative was the second one. Table 4.2 presents the values of the criteria for every one of the three alternative actions. Furthermore, it updates the information stored in the user model maintained on the UM-Server.

Table 4.2. The values of the criteria based on the information stored in UM-Server user model.

	f	e	s	d	g
J1	0.2	0.3	0.3	0.5	0.4
J2	0.7	0.3	0.9	0.5	1.0
J3	0.5	0.3	0.6	0.5	0.4

The same user also interacts with MBIFM. The user has emptied the folder 'C:\Courses\SoftwareEng1' and moved its content to the folder 'C:\Courses\Old' that she keeps all files about her old courses. Then she selects the folder 'C:\Courses\SoftwareEng2' and issues first the command 'copy' and then the command 'delete'. However, the system finds the particular action as not intended as the particular folder contains many subfolders and files that will be lost and the system generates alternative actions.

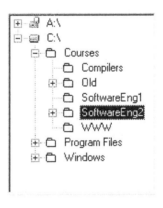

Fig. 4.2. The user's initial file store state

The alternative actions that are generated by the system are the following:

G1. delete(C:\Courses\SoftwareEng1\)

G2. delete(C:\Courses\WWW\)

G3. cut(C:\Courses\SoftwareEng1\)

The first alternative action is based on the assumption that the user has tangled up neighbouring and similarly named objects. Similarly, the second alternative supposes that the user has tangled up neighbouring objects whereas the last alternative is based on the assumption that the user wanted to issue a different command. However, the user model that is maintained centrally on UM-Server reveals that the user is an expert and, therefore it is rather unlikely to have been mistaken in selecting the correct command. Furthermore, it has proven from the user's interaction with MI-Mailer and INTATU that the particular user is prone to accidental slips and selecting the wrong object in a graphical representation. Using the values of the criteria as these were formed after the user's last interaction with MI-Mailer and INTATU (Table 4.3), the Web Service selects the first alternative action and sends it to MBIFM to propose it to the user. The way that the values of the criteria are acquired by UM-Server and then used by INTATU, MI-Mailer and MBIFM in order to select the best alternative is presented in the next sections.

Table 4.3. The values of the criteria based on the information stored in UM-Server user model.

	f	e	s	d	g
G1	0.8	0.3	0.9	0.5	1.0
G2	0.2	0.3	0.3	0.5	0.4
G3	0.1	0.4	0.4	0.6	0.0

4.5. User Modelling based on Web Services

When a user logs on INTATU, MBIFM or MI-Mailer then the UM-Server is responsible for locating the user model and returning this information to the application that requested the information. The operation of UM-Server is quite simple. The application makes a request to the Server and the Server returns an XML document containing the response to the application's request. Based on this, the system sends the username and password of the user to the UM-Server and the latter is responsible for finding the user model and sending this information to the client that requested it. The user model is updated with information gathered locally, through the user's interaction with the application. Finally, the information acquired is sent back to the UM-Server so that the user model is updated. In this way, UM-Server keeps track of the intentions and the possible confusions of each individual user. This information is available to the client application irrespective of the computer where it is running.

4.5.1. UM-Server's Architecture

In every one of the systems presented above (MBIFM,MI-Mailer, INTATU), most of the user models' manipulation is handled by UM-Server and the only data sent back to the client application are query results. Then the client application can further process the results in order to calculate the values of the attributes and, finally, adapt its interaction with each individual user. The main goal here is to ensure basic interoperability. This may be achieved by using standard communication protocols. Therefore, the UM-Server architecture is based on Web Services, which are based on standard protocols. The primary standards involved in this approach are:

- XML (eXtensible Markup Language) to handle the data transportation,
- SOAP (Simple Object Access Protocol) to handle communications,
- WSDL (Web Services Description Language) to provide the metadata necessary to run the service, and
- UDDI (Universal Description Discovery and Integration) to register services on Internet servers and thus make them public.

Communication between a client and UM-Server takes place in terms of Web Services protocols. The client application makes specific SOAP calls (under HTTP), which contain requests to the Web server. The Communication module retrieves and handles each such call, recognising it in the URL's structure. After resolving the service's part of the URL it received, the Communication Module passes on the rest of the string to the DB Module or the User Modelling Module in order to process the request and form the response. The response is sent again to the Communication Module, which encodes it in XML and returns it to the caller (client).

The architecture of UM-Server is illustrated in Figure 4.3. The functionality of each module of UM Server is presented below:

- The Communication module handles all the Web Service messages. In particular, it handles authentication requests, profile requests, profile update requests, creation requests and deletion requests. Furthermore, it is responsible for formatting the response in XML and sending it to

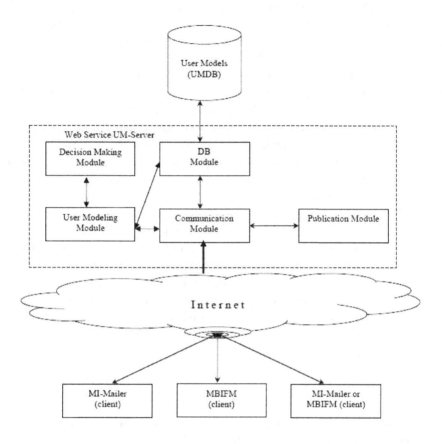

Fig. 4.3. The architecture of UM-Server

the client application.
- The database module (DB module) implements all the functions necessary to create, update and delete profiles, as well as to authenticate a user. In order to perform these functions, the DB module interacts with the database of user models (UMDB).
- The User Modelling Module processes all the information acquired by

the UMDB, the Decision Making Module or the client application in order to draw basic inferences about the user or form the response to the call made by the client application.
- The Decision Making Module processes all information acquired by the User Modelling Module and applies SAW in order to select the alternative action to be proposed to the user by the system. More specifically, the User Modelling Modules provides a set of alternatives actions and the values of criteria for each alternative action and the Decision Making Module applies SAW to select the best one.
- The Publication Module handles the interface publication demands. It presents a description in HTML language for a developer that wishes to use such a profile and handles the WSDL requests concerning the Web Service. It handles all the details necessary for the interaction with the service, including message formats, transport protocols and location. More specifically, the information concerning the operation of the Web Service contains an identification of the service, a complete list of the functions provided together with their arguments and return values. The interface hides the implementation details of the service, allowing it to be used independently of the hardware or software platform on which it is implemented on and also independently of the programming language which it is written in.

4.5.2. UM-Server's Operation

In order to provide intelligent help or assistance during e-Learning, IN-TATU, MBIFM and MI-Mailer monitor all users' actions and reason about them. The information collected by this process is stored in a detailed user model, which is located on a web server. Therefore, every time the user interacts with one of the two systems, the client makes a specific SOAP call, which contains a request about the particular user model, to the Web server. In order to authenticate the user, this call contains the username and the password given by the user during his/her interaction with the mailing or the file manipulation system. The retrieval of this request is the main task of the Communication Module. After resolving the service's part of the URL it received, the Communication Module passes on the rest of the string to DB Module. The DB Module authenticates the user and finds his/her user model.

In case the DB Module succeeds in finding the user model, the information acquired is further processed by the User Modeling Module and the result is sent to the Communication Module. The Communication Module generates a response in XML, which contains information about the values of the attributes for the user interacting with the client application that made the call, and returns it to the caller (client). In this case the caller could be any of the two applications discussed in this paper or even a third one that uses a similar user model.

In case the user does not exist in the User Models Database (UMDB), the DB Module informs the Communication Module for its failure to find the particular user model. The Communication Module generates a response in XML that informs the caller (client) that there is no available information about the user's preferences and characteristics. Therefore, the client asks the user a number of questions about his/her previous knowledge of e-mail clients, file manipulation programs, operating systems, his/her Internet experiences etc. The information collected by this process is sent to UM-Server as a specific SOAP call. The Communication Module that receives the request, sends it to the User Modelling Module. As the User Modelling Module maintains a library of models for every group of users, it is responsible for retrieving the information about the user and deciding which stereotype should be activated.

In UM-Server, users are classified into one of three major stereotype classes according to their level of expertise, namely, novice, intermediate and expert. Each one of these classes represents an increasing mastery in the use of the particular e-mailing system. Such a classification was considered crucial because it would enable the system to have a first view of the usual errors and misconceptions of a user, belonging to a group. For example, novice users are usually prone to mistakes due to erroneous command selection or erroneous command execution whereas expert users usually make mistakes due to carelessness. Therefore, another classification that was considered important was dividing users into two groups, careless and careful.

All default assumptions in stereotypes used in UM-Server, give information about the errors that users belonging to each category usually make. Information about each error is expressed by using the values of the attributes f and e. More specifically, the value of f is used for showing how often users belonging to a certain group make a particular error. Another piece of information that can be derived from a stereotype is how the success of the user in executing each command, which is expressed by the value of the attribute e.

As the user interacts with the client application (INTATU, MBIFM or

MI-Mailer), it collects more and more information about him/her and stores it in an individual user model. Therefore, every time the user interacts with the client, the information collected is sent to the UM-Server, which makes all the essential work in order to update the user model. In this way, the UM-Server keeps track of intentions and possible confusions of each individual user using the values of the attributes. This information is available to any client application irrespective of the computer where it is running.

4.6. Multi-Attribute Decision Making on the Server side

The value of the attribute f that refers to the frequency of an error is calculated by dividing the times a particular user has made an error by his/her total errors. Such information is provided by the user model that is maintained centrally on the Server. This user model also provides information for calculating the value of the attribute e that represents the percentage of the wrong executions of a command in the number of total executions of the particular command. The particular attribute is calculated by dividing the times the user has made an error in the execution of a command by the total number of the command's execution. The degree of relevance to the user's goals, on the other hand, is estimated by taking into account the information about the user's goals and plans that is stored in the individual short term user model. If the alternative action that is evaluated results in the completion or continuation of a plan then the value of the particular attribute is 1 otherwise, its value is lower. Finally, the values of the other two attributes, s and d, are calculated by taking into account the information that is stored in the knowledge representation component of the system. For example, the degree of difficulty of each command is a prefixed value that is maintained constant for all users.

However, the above mentioned attributes are not equally important in the reasoning process of the human experts. For this purpose, an empirical study was conducted in order to identify how important each attribute is in the reasoning process of human experts. More specifically, 16 human advisors were asked to rank the five attributes with respect to how important these attributes were in their reasoning process. However, SAW does not propose a standard procedure for setting a rating scale for attributes weights. Several researchers have used different scale rating. For example use a scale from 1 (least desirable) to 9 (most desirable) for six different attributes.

In view of the above, a scale from 1 to 5 is proposed for rating the attributes in this empirical study. More specifically, every one of the 16 human experts was asked to assign one score of the set of scores (1, 2, 3, 4, 5) to each one of the four attributes and not the same one to two different attributes. The sum of scores of the elements of the set of scores was 15 (1+2+3+4+5=15). For example, a human expert could assign the score 5 on the degree of relevance to the user's goals (attribute g), the score 4 on the frequency of an error (attribute f), the score 3 on the percentage of the wrong executions of a command in the number of total executions of the particular command (attribute e), the score 2 on attribute s (similarity of an alternative action with the actual action issued by the user) and the score 1 on the attribute d (difficulty of a command).

As soon as the scores of all human experts were collected, they were used to calculate the weights of the attributes. The scores assigned to each attribute by each human expert were summed up and then divided by the sum of scores of all attributes (15 * 16 human experts = 240). In this way the sum of all weights could be equal to 1.

As a result, the calculated weights for the attributes were the following:

- The weight for the degree of similarity (s): $w_s = \frac{75}{240} = 0.313$
- The weight for the frequency of an error (f): $w_f = \frac{39}{240} = 0.163$
- The weight for percentage of the wrong executions of a command in the number of total executions of the particular command (e): $w_e = \frac{37}{240} = 0.154$
- The weight for the degree of difficulty of a command (d): $w_d = \frac{27}{240} = 0.113$
- The weight for the degree of elevance to the user's goals (g): $w_g = \frac{62}{240} = 0.258$

The empirical study for calculating the weights of the attributes is only conducted once, but the values of the attributes are calculated every time one alternative action is to be evaluated. Therefore, as soon as the decisions about the calculation of the values of the attributes for every alternative action have been made, SAW may be incorporated in the Web Service.

In SAW, the alternative actions are ranked by the values of a multi-attribute function that is calculated for each alternative action as a linear combination of the values of the n attributes. So the client application (INTATU, MBIFM or MI-Mailer) gathers the values of all attributes and send it to the Web Service, which is responsible for calculating the values of the multi- attributes utility function U for each alternative action generated by the system. More specifically, the function U takes its values as a linear combination of the values of the five attributes described in the previous section:

$$U_{SAW}(X_i) = w_e e_i + w_v v_i + w_c c_i + w_d d_i, \qquad (4.1)$$

where X_i is the evaluated alternative, w_e, w_v, w_c, w_d are the weights of the attributes and e_i, v_i, c_i, d_i are the values of the attributes for the i alternative. As the weights of the attributes are already known the multi-attribute utility function is transformed to be:

$$U_{SAW}(X_i) = 0.313 s_i + 0.258 g_i + 0.163 f_i + 0.154 e_i + 0.113 d_i. \qquad (4.2)$$

The particular formula is used for selecting the best alternative action to be proposed to the user. For the example interactions presented in section 4, the particular formula was used in the Web Service in order to select the best alternative action irrespective of the application that the user interacted with. More specifically, in the case of the undesired action in the user's interaction of INTATU, the Web Service uses the values of the criteria presented in Table 4.1 for estimating the value of using formula (4.2): $U_{SAW}(T1) = 0.685$, $U_{SAW}(T2) = 0.404$. Taking into account the values of the multi-attribute utility function, INTATU selects the first alternative action $T1$ to present it to the user.

Similarly, in the case of MI-Mailer, the Web Service used the values of the attributes presented in Table 4.2 for applying formula (4.2) in every one of the three alternative actions: $U_{SAW}(J1) = 0.33$, $U_{SAW}(J2) = 0.76$ $U_{SAW}(J3) = 0.47$. As a result, MI-Mailer proposed the second alternative that had the highest value of the multi-attribute utility function. Finally, when the user interacted with MBIFM, the Web Service used the values of the attributes presented in Table 4.2 for applying formula (4.2) in every one of the three alternative actions:$U_{SAW}(G1) = 0.77$, $U_{SAW}(G2) = 0.33$ $U_{SAW}(G3) = 0.27$. The multi-attribute utility function is maximised in the case of the first alternative, and, therefore, $G1$ is selected to be presented to the user.

4.7. Conclusions

In this paper, we have shown how applications from different domains can share a common user model and adapt their interaction to each user. More specifically, the paper presents an e-learning application for health issues and two intelligent user interfaces, one for file manipulation and one for e-mailing. Each one of the above mentioned applications gathers information about its users and stores it in a common user model that is maintained in a user model server, called UM-Server. UM-Server maintains a central database of all user models and allows the client applications to access this information from virtually everywhere.

UM-Server uses Web Services. The technology of Web Services enables the creation of a highly re-usable user model server as it offers advantages to the developers of applications to create applications that are interoperable and highly reusable. However, the reusability of Web Services depends on the functionality that has been embodied in the Web Service and thus it poses a problem for the application developers to solve.

For this purpose, in UM-Server the functionality of user modelling is performed on the server side (UM-Server) and, thus, can be used by many different applications. More specifically, UM-Server performs user modelling based on a multi-criteria decision making theory called SAW. In this way we have addressed the major problem caused by the need of construction from scratch of user models that incorporate a multi-criteria decision making theory for various applications. Moreover, the fact that we have incorporated a multi-criteria decision making theory into the reasoning of a user model server, provides a reusable user modelling mechanism that is not based on ad-hoc methods but rather it adapts and implements a well-established multi-criteria theory for the purposes of user modelling.

Appendix A. Multi-Attribute Decision Making

A multi-attribute decision problem is a situation in which, having defined a set A of actions and a consistent family F of n attributes g_1, g_2, ..., g_n ($n \geq 3$) on A, one wishes to rank the actions of A from best to worst and determine a subset of actions considered to be the best with respect to F.[18] According to[19] there are three steps in utilising a decision making technique that involves numerical analysis of alternatives:

(1) Determining the relevant attributes and alternatives.
(2) Attaching numerical measures to the relative importance of the attributes and to the impacts of the alternatives on these attributes.
(3) Processing the numerical values to determine a ranking of each alternative.

The determination of the relevant attributes and their relative importance is made at the early stages of the software life-cycle and is performed by the developer or is based on an empirical study which may involve experts in the domain. However, decision making techniques mainly focus on step 3. The Simple Additive Weighting (SAW)[10,11] method is probably the best known and most widely used decision making method. SAW consists of two basic steps:

(1) **Scale the values of the n attributes to make them comparable.** There are cases where the values of some attributes take their values in [0,1] whereas there are others that take their values in [0,1000]. Such values are not easily comparable. A solution to this problem is given by transforming the values of attributes in such a way that they are in the same interval.

(2) **Sum up the values of the n attributes for each alternative.** As soon as the weights and the values of the n attributes have been defined, the value of a multi-attribute function is calculated for each alternative as a linear combination of the values of the n attributes.

Scaling the values of the n attributes to make them comparable is the first step in applying the SAW method. However, the way that the values of the attributes should be scaled up is not defined by SAW and, therefore, different approaches have been proposed in order to overcome this problem. The one proposed by[20] is a very good one as it transforms all values so that they take their values in the interval [0,1] and has been applied in solving decision making problems in computer science. More specifically,[20] uses the scaling factor of formula A.1 in order to normalise the values of attributes such as understandability, extent, availability, time and price.

$$x_{ij} = \frac{d_{ij} - d_j^{min}}{d_j^{max} - d_j^{min}}, \quad (A.1)$$

where d_{ij} is the old value and x_{ij} is the transformed value of the j attribute for the i alternative and are the maximum and minimum values of the j attribute for all the alternatives. Using this scaling, the values of all attributes are in the interval [0,1].

The SAW approach consists of translating a decision problem into the optimisation of some multi-attribute utility function U defined on A. The decision maker estimates the value of function for every alternative X_j and selects the one with the highest value. The multi-attribute utility function U can be calculated in the SAW method as a linear combination of the values of the n attributes:

$$U(X_j) = \sum_{i=1}^{n} w_i x_{ij}, \qquad (A.2)$$

where X_j is one alternative and x_{ij} is the value of the i attribute for the X_j alternative.

References

1. E. Rich, User modelling via stereotypes, *Cognitive Science.* **3**, 329–354, (1979).
2. E. Rich, Users are individuals: individualizing user models, *International Journal of Human-Computer Studies.* **51**, 323–338, (1999).
3. J. Kay, *Student Models and Scrutability*, In ed. K. V. G. Gautier, C. Frasson, Intelligent Tutoring Systems (Proceedings of the 5th International Conference on Intelligent Tutoring Systems), pp. 19–30. Lecture Notes in Computer Science. Springer Verlag, Berlin, (2000).
4. L. Ardissono and A. Goy, *Tailoring the Interaction With Users in Electronic Shops*, In User Modeling: Proceedings of the Seventh International Conference. Lecture Notes in Artificial Intelligence. Springer Verlag, (1999).
5. D. N. Chin, *KNOME Modeling What the User Knows in UC*, In eds. A. Kobsa and W. Wahlster, User Models in Dialog Systems, pp. 74–107. (1989).
6. M. L. J. M. J. M. R. Wilensky, D.N. Chin and D. Wu, The berkeley unix consultant project, *Artificial Intelligence Review, Intelligent Help Systems For Unix, volume =.*
7. B. T. T. K. B. T. R. Kaschek, K.D. Schewe. Learner typing for electronic learning systems. In *Proceedings of the IEEE International Conference on Advanced Learning Technologies*, (2004).
8. B. K. J. Kay and P. Lauder, *Personis: A Server for User Models*, In eds. P. B. P. De Bra and R. Conejo, AH 2002, vol. 2347, LNCS, pp. 203–212. Springer Verlang Berlin Heidelberg, (2002).
9. J. Fink and A. Kobsa, A review and analysis of commercial user modeling servers for personalization on the world wide web, *User Modeling and User-Adapted Interaction.* **10**, 209–249, (2000).
10. P. C. Fishburn, Additive utilities with incomplete product set: Applications to priorities and assignments, *Operations Research.* (1967).
11. C. L. Hwang and K. Yoon, *Multiple Attribute Decision Making: Methods and Applications*, In Lecture Notes in Economics and Mathematical Systems, vol. 186. Springer Verlag.
12. K. Kabassi and M. Virvou. Software engineering aspects of an intelligent user interface based on multi criteria decision making. In *The First International Workshop on Web Personalization, Recommender Systems and Intelligent User Interfaces (WPRSIUI 2005)*, Reading, U.K (Oct., 2005).
13. A. Sheikhzadeh and P. Ehlermann, Atheromatosis disease of the thoracic aorta and systemic embolism clinical picture and therapeutic challenge, *Zeitschrift fur kardiologie.* **93**, 1–17, (2004).

14. S. Delisle and B. Moulin, User interfaces and help systems: From helplessness to intelligent assistance, *Artificial Intelligence Review.* **18**, 117–157, (2002).
15. M. Virvou and K. Kabassi, *An Empirical Study Concerning Graphical User Interfaces that Manipulate Files*, In eds. J. Bourdeau and R. Heller, *Proceedings of ED-MEDIA 2000, World Conference on Educational Multimedia, Hypermedia & Telecommunications*, pp. 1724–1726. AACE, Charlottesville VA, (2000).
16. K. Kabassi and M. Virvou, A knowledge-based software life-cycle framework for the incorporation of multi-criteria analysis in intelligent user interfaces, *IEEE Transactions on Knowledge and Data Engineering.* **18**, 1–13, (2006).
17. M. V. K. Kabassi and G. Tsihrintzis, In ed. H. I. M. T. for Atheromatosis, *Knowledge-Based Intelligent Information and Engineering Systems KES 2006*. Lecture Notes in Artificial Intelligence, subseries of Lecture Notes in Computer Science. Springer, Berlin, (2006).
18. P. Vincke, *Multicriteria Decision-Aid.* (Wiley, 1992).
19. E. Triantaphyllou and S. Mann, An examination of the effectiveness of four multi-dimensional decision-making methods: A decision-making paradox, *International Journal of Decision Support Systems.* **5**, 303–312, (1989).
20. F. Naumann. Data fusion and data quality. In *Proceedings of the New Techniques and Technologies for Statistics*, (1998).

BIOGRAPHIES

Katerina Kabassi is a Professor of Technological Applications in the Department of Ecology and the Environment, Technological Educational Institute of the Ionian Islands (Greece). She received a first degree in Informatics (1999) and a D.Phil. (2003) from the Department of Information, University of Piraeus (Greece). She has authored over 39 articles, which have been published in international journals, books and conferences. She has served as a member of Program Committees and/or reviewer of International conferences. Her current research interests are in the areas of Knowledge based Software Engineering, Human Computer Interaction, Personalization Systems, Multi-Criteria Decision Making, User Modeling, Web-based Information Systems and Educational Software.

Maria Virvou is an Associate Professor in the Department of Informatics, University of Piraeus, Greece. She received a degree in mathematics from the University of Athens, Greece (1986), a M.Sc. (Master of Science) in computer science from the University of London (University College London), UK (1987) and a D.Phil. from the School of Cognitive and Computing Sciences of the University of Sussex, UK (1993). She is the sole author of three computer science books. She has authored or co-authored over 150 articles, which have been published in international journals, books and conferences. She has served as a member of Program Committees and/or reviewer of international journals and conferences. She has supervised and is currently supervising 12 Ph.D.s She has served and is serving as the project leader and/or project member in 15 R&D projects in the areas of e-learning, computer science and information systems.

Her research interests are in the areas of web-based information systems, knowledge-based human computer interaction, personalization systems, software engineering, e-learning, e-services and m-services.

George A. Tsihrintzis received the Diploma in electrical engineering from the National Technical University of Athens (with honors) in 1987 and the M.Sc. and Ph.D. degrees in electrical engineering from Northeastern University in 1988 and 1992, respectively. He is currently an Associate Professor in the Department of Informatics, The University of Piraeus, Greece.

He has authored or co-authored over 150 articles in these areas, which have appeared in international journals and conferences and has served as the project leader in several R&D projects. He is a member of the IEEE.

His current research interests include pattern recognition, decision theory, and statistical signal processing, and their applications in user modeling, intelligent software systems, human-computer interaction and information retrieval.

PART 2
Collaborative Filtering

Chapter 5

Experimental Analysis of Multiattribute Utility Collaborative Filtering on a Synthetic Data Set

Nikos Manouselis* and Constantina Costopoulou[†]

Informatics Laboratory, Division of Informatics, Mathematics & Statistics Agricultural University of Athens, 75 Iera Odos str., 118 55 Athens, Greece
nikosm@ieee.org
[†]*tina@aua.gr*

Recommender systems have already been engaging multiple criteria for the production of recommendations. Such systems, referred to as multi-criteria recommenders, early demonstrated the potential of applying Multi-Criteria Decision Making (MCDM) methods to facilitate recommendation in numerous application domains. On the other hand, systematic implementation and testing of multi-criteria recommender systems in the context of real-life applications still remains rather limited. Previous studies dealing with the evaluation of recommender systems have outlined the importance of carrying out careful testing and parameterization of a recommender system, before it is actually deployed in a real setting. In this paper, the experimental analysis of several design options for three proposed multi-attribute utility collaborative filtering algorithms is presented. The data set used is synthetic, with multi-criteria evaluations that have been created using an appropriate simulation environment. This synthetic data set tries to mimic the evaluations that are expected to be collected from users in a particular application setting. The aim of the experiment is to demonstrate how a synthetic data set may be created and used to facilitate the study and selection of an appropriate recommendation algorithm, in the case that multi-criteria evaluations from real users are not available.

5.1. Introduction

The area of recommender systems attracts high research interest due to its challenging open issues.[1] Nowadays, there is an abundance of real-life applications of recommender systems in the Web, which may help Internet users to deal with information overload by providing personalized recommendations regarding online content and services.[2] The application domains range from recommendation of commercial products such as books, CDs and movies, to recommendation of more complex items such as quality methods and instruments. Early recommender systems were based on the notion of collaborative filtering, and have been defined as systems that "... help people make choices based on the opinions of other people".[3] With time, the term *recommender systems* has prevailed over the term *collaborative filtering systems*.[4] It evolved to cover "... any system that produces individualized recommendations as output or has the effect of guiding the user in a personalized way to interesting or useful objects in a large space of possible options".[5]

In a recommender system, the items of interest and the user preferences are represented in various forms, which may involve one or more variables. Particularly in systems where recommendations are based on the opinion of others, it is crucial to incorporate the multiple criteria that affect the users' opinions into the recommendation problem. Several recommender systems have already been engaging multiple criteria for the production of recommendations. Such systems, referred to as multicriteria recommenders, demonstrated early the potential of applying *multi-criteria decision making* or MCDM methods to facilitate recommendation in numerous application domains, such as movie recommendation,[6] restaurant recommendation,[7] product recommendation,[8] and others.[9]

In their recent survey of the state-of-the-art in the field of recommender systems, Adomavicius and Tuzhilin[1] stated that MCDM methods may have been extensively studied in the Operations Research domain, but their application in recommender systems has yet to be systematically explored. An observation supporting their statement is that systematic implementation and testing of multicriteria recommender systems in the context of real-life applications still remains rather limited.[1,10,11] This indicates that the evaluation of multicriteria recommender systems is not in line with the conclusion of previous studies dealing with recommender systems' evaluation. These studies (e.g.[12–15] have outlined the importance of carrying out careful testing and parameterization of a recommender system, before it is finally deployed in a real setting.

Towards this direction, this paper experimentally investigates various design choices in a multicriteria recommender system, in order to support neighborhood-based collaborative filtering in a particular application context. More specifically, Sec. 5.2 describes how collaborative filtering may be modeled using multi-attribute utility theory or MAUT principles.[16] In Sec.

5.3, a classic neighborhood-based algorithm for single-criterion collaborative filtering is extended to support multi-attribute collaborative filtering.[16] Then, a classic neighborhood-based algorithm for single-criterion collaborative filtering is extended to support multicriteria collaborative filtering. Three different MAUT-based techniques for calculating the similarity between neighbors are considered, leading to three multiattribute utility algorithms. Following the guidelines of related literature, various design options for each algorithm are considered. In Sec. 5.4, the proposed algorithms are experimentally evaluated for potential implementation in an examined application context: multi-attribute recommendation of electronic markets (e-markets) to online customers. For this purpose, a synthetic data set with multi-criteria evaluations is created, using an appropriate simulation environment. Using the created data set, several design options are explored for each algorithm. Using the collected data set, several design options are explored for each algorithm. In Sec. 5.5, a discussion of the benefits and shortcomings of the proposed approach is provided. Finally, Sec. 5.6 outlines the conclusions of this study and directions for future research.

5.2. MAUT Collaborative Filtering

In related research, the problem of recommendation has been identified as the way to help individuals in a community to find the information or products that are most likely to be interesting to them or to be relevant to their needs.[17] It has been further refined to the problem (i) of predicting whether a particular user will like a particular item (prediction problem), or (ii) of identifying a set of N items that will be of interest to a certain user (top-N recommendation problem).[13] Therefore, the recommendation problem can be formulated as follows:[1] let C be the set of all users and S the set of all possible items that can be recommended. We define as $U^c(s)$ a utility function $U^c(s) : C \times S \to \Re^+$ that measures the appropriateness of recommending an item s to user c. It is assumed that this function is not known for the whole $C \times S$ space but only on some subset of it. Therefore, in the context of recommendation, we want for each user $c \in C$ to be able to:

- estimate (or approach) the utility function $U^c(s)$ for an item s of the space S for which $U^c(s)$ is not yet known, or
- choose a set of items $S' \subseteq S$ that will maximize $U^c(s)$:

$$\forall c \in C, s = \max_{s \in S'} U^c(s). \qquad (5.1)$$

In most recommender systems, the utility function $U^c(s)$ usually considers one attribute of an item, e.g. its overall evaluation or *rating*. Nevertheless, utility may also involve more than one attributes of an item. The recommendation problem may therefore be viewed under the prism of MCDM.[18–20] An extensive review and analysis of how multi-criteria recommender systems support the users decision has been performed and is presented elsewhere[11] In this paper, we focus on Value-Focused models, and more specifically MAUT ones. Several MAUT recommender systems have already been introduced in related literature.[7,21–24] In general, Value-Focused models have already been applied in recommender systems, such as the listed MAUT approaches or other approaches that may be found in the literature.[25–27]

Multi-criteria recommender systems have the advantage that they consider more than one criterion that may affect the potential users decision, in order to make a recommendation. However, most current proposals remain at a design or prototyping stage of development. Until today, the systematic design, implementation, and evaluation of multi-criteria recommenders in the context of real-life applications is limited (e.g.[23]). In addition, the systematic evaluation of multi-criteria recommenders requires their experimental investigation in the context of the particular application domains, using data sets with multi-criteria evaluations.[10]

Collaborative recommendation (or collaborative filtering) takes place when a user is recommended items that people with similar tastes and preferences liked in the past.[1] Collaborative filtering systems predict a user's interest in new items based on the recommendations of other people with similar interests. Instead of performing content indexing or content analysis, collaborative filtering systems rely entirely on interest ratings from the members of a participating community.[15] The problem of automated collaborative filtering is to predict how well a user will like an item that he has not rated (also called "evaluated" in the rest of this paper), given a set of historical ratings for this and other items from a community of users.[15] In single-attribute (or single criterion) collaborative filtering, the problem space can be formulated as a matrix of users versus items (or *user-rating* matrix), with each cell storing a user's rating on a specific item. Under this formulation, the problem refers to predicting the values for specific empty cells (i.e. predict a user's rating for an item). Following the notation of Sec. 5.2, it can be said that collaborative filtering aims to predict the utility of items for a particular user (called *active user*), based on the items previously evaluated by other users.[1] That is, the utility $U^a(s)$ of item s for the active user $a \in C$ is estimated based on the utilities $U^c(s)$ assigned to item s by those users $c \in C$ who are "similar" to user a. For classic, single-attribute collaborative filtering, this corresponds to the prediction of the rating $U^a(s) = r_{a,s}$, according to the ratings $U^c(s) = r_{c,s}$ provided by the users $c \in C$ who are "similar" to user a.

Engaging MAUT,[16] the recommendation problem in collaborative filtering systems may be defined as a decision problem with multiple variables (called *multiattribute utility collaborative filtering*), which may be modeled in the following manner. The multiple attributes describing an item s are defined as a set of criteria upon which a user evaluates the item. The utility function $U^c(s)$ is then referred to as the *total utility* of an item s, which is calculated by synthesizing the *partial utilities* of the item s on each one of the criteria. The criteria are independent, nondecreasing real-valued functions, defined on S as $g_i : S \to \Re$, where $g_i(s)$ is the evaluation of the item s on the ith criterion ($i = 1, \ldots, n$). Thus, the multicriteria evaluation of an item $s \in S$ is given as a vector $\underline{g}(s) = [g_1(s), g_2(s), \ldots, g_n(s)]$. The global preference model is formulated as an additive value function, where an importance weight is associated with each evaluation criterion. Assuming that there is no uncertainty during the decision making, the total utility of an item $s \in$ for a user $c \in C$ can be expressed as:

$$U^c(s) = \sum_{i=1}^{n} u_i^c(s) = \sum_{i=1}^{n} w_i^c g_i^c(s), \qquad (5.2)$$

where $u_i^c(s)$ is the partial utility function of the item s on criterion g_i for the user c, $g_i^c(s)$ is the evaluation that user c has given to the item s on criterion g_i and w_i^c is the weight indicating the importance of criterion g_i for the particular user c, with:

$$\sum_{i=1}^{n} w_i^c = 1. \qquad (5.3)$$

The linear function of the total utility function is the simplest and most popular form of an additive value function. Other forms that could be used include an ideal point model, dependencies and correlations, as well as diminishing utility forms (see[9]).

For each user $c \in C$ that has evaluated an item $s \in S$, this evaluation is given as a vector $\underline{g}^c(s) = [g_1^c(s), \ldots, g_n^c(s)]$, and there is also a set of importance weights $\underline{w}^c = [w_1^c, \ldots, w_n^c]$ that are associated with the n criteria.

In the remainder of this paper, the evaluations $g_i^c(s)$ are referred to as the *evaluations* of user c, and the weights $w_i^c (i = 1, \ldots, n)$ as the *properties* of user c.

5.3. MAUT Algorithms for Collaborative Filtering

The goal of the collaborative filtering system is to provide to the active user $a \in C$, either an estimation of the total utility for a particular target item s that he has not previously evaluated, or a *ranking* of a subset of items $S'' \subseteq S$. For the items in S'' that the active user a has not evaluated yet, this corresponds again to the prediction of the utility $U^a(s)$, for each item $s \in S''$ that this user has not evaluated. Thus, we will address both goals in a similar manner, by calculating the prediction of $U^a(s)$. To calculate this prediction, we engage a *neighborhood-based* collaborative filtering algorithm.

Neighborhood-based algorithms are the most prevalent approaches for single-criterion collaborative filtering.[15,28] They belong to the category of memory-based ones, and they have their roots in instance-based learning (IBL) techniques that are very popular in machine learning applications.[29] The nearest neighbor algorithm is one of the most straightforward IBL ones.[30] During generalization, IBL algorithms use a distance function to determine how close a new instance is to each stored instance, and use the nearest instance or instances to predict the target.[28] There are several proposed approaches for neighborhood-based collaborative filtering (e.g.[12,15,28,31]). These approaches engage various methods and techniques at each stage of a neighborhood-based algorithm, in order to acquire an accurate prediction. To design the multiattribute algorithms, we build upon a number of stages of single-attribute neighborhood-based algorithms, as they have been identified by Herlocker *et al.*[15] and extended by other researchers:

- *Stage A — Similarity Calculation*: this is the core stage of the algorithm, where the similarity between the examined user (active user) and the rest users is calculated.
- *Stage B — Feature Weighting*: the engagement of a feature weighting method further weights similarity according to the characteristics of each examined user or some heuristic rules.
- *Stage C — Neighborhood Formation/Selection*: it refers to the selection of the set of users to be considered for producing the prediction.

- *Stage D — Combining Ratings for Prediction*: the final stage, normalizing the ratings that the users in the neighborhood have provided for the unknown item, and using some method to combine them in order to predict its utility for the active user.

Neighborhood-based algorithms therefore create a neighborhood $D \subseteq C$ of m users that have similar preferences with the active user and who have previously evaluated the target item s, and calculate the prediction of $U^a(s)$ according to how the users in the neighborhood have evaluated s. That is, if m is the number of users in the neighborhood D, the recommendation algorithm will predict $U^a(s)$ according to the m utilities $U^d(s)$ of this item for each neighbor $d \in D$. In the following, we examine three different algorithms for MAUT-based collaborative filtering. Each algorithm formulates the neighborhood D based on a different notion of how "similar preferences" can be measured. Other algorithms can also be considered, according to how preference similarity is measured.[32]

5.3.1. *Proposed algorithms*

5.3.1.1. *Similarity per priority (PW) algorithm*

This algorithm is based on including in the neighborhood $D \subseteq C$ users that have similar priorities to the properties w_i^a ($i = 1, \ldots, n$) of the active user. That is, it bases the recommendation on the opinion of users that assign similar importance to each evaluation criterion when selecting an item. The various design options for the PW algorithm are illustrated in Table 5.1 (a detailed description may be found in[11]). The options examined for similarity calculation measure the distance between the vector of the properties of the active user a and the vector of the properties of user c.

5.3.1.2. *Similarity per evaluation (PG) algorithm*

This algorithm calculates the prediction of the total utility $U^a(s)$ of a target item $s \in S$, by calculating the n predictions of how the active user would evaluate s upon each criterion g_i ($i = 1, \ldots, n$) in separate, and then synthesizing these predictions into a total utility value. This algorithm is in line with the proposed approach for multidimensional recommenders presented by Adomavicius et al.,[33] where n-dimensional recommendations are calculated by synthesizing the *User* × *Item* recommendations upon each one of the n dimensions. The algorithm creates n neighborhoods $D_i \subseteq C$,

Table 5.1. Design options for the PW algorithm.

Algorithm Stage	Design Options	Description
Similarity Calculation	Euclidian distance	$\text{sim}(a,c) = 1 - \sqrt{\dfrac{\sum_{i=1}^{n} f_i^2 (w_i^a - w_i^c)^2}{\sum_{i=1}^{n} f_i^2}}$
	Vector/Cosine similarity	$\text{sim}(a,c) = \dfrac{\sum_{i=1}^{n} f_i^2 (w_i^a \cdot w_i^c)}{\sqrt{\sum_{i=1}^{n} f_i^2 \cdot (w_i^a)^2} \times \sqrt{\sum_{i=1}^{n} f_i^2 \cdot (w_i^c)^2}}$
	Pearson correlation	$\text{sim}(a,c) = \dfrac{\sum_{i=1}^{n} f_i^2 (w_i^a - \overline{w^a})(w_i^c - \overline{w^c})}{\sqrt{\sum_{i=1}^{n} f_i^2 (w_i^a - \overline{w^a})^2 \times \sum_{i=1}^{n} f_i^2 (w_i^c - \overline{w^c})^2}}$ where $\overline{w^a}$ is the mean value of the priorities of the active user, and $\overline{w^c}$ the mean value of the priorities of the other user.
Feature Weighting	None	$f_i = 1$
	Inverse user frequency	$f_i = \log \dfrac{c_{\max}}{c_i}$ where c_i is the number of users that have provided a priority on criterion i, and $c_{\max}$ is the total number of users in the system. Since we assume that after the data processing stage all users will provide priorities on all criteria, this option is equivalent to the previous one ($f_i = 1$).
	Entropy	$f_i = \dfrac{H_i}{H_{i,\max}}$ where $H_i = -\sum_{i=\min(g_i)}^{\max(g)_i} p_{j,i} \cdot \log_2 p_{j,i}$, H_i is the entropy of priorities on criterion i, $p_{j,i}$ is the probability of priorities on criterion i to take the value j (distribution of priorities upon scales [min$(w_i), \ldots, \max(w_i)$]), and $H_{i,\max}$ represents the maximum entropy which assumes that the distributions over all scales of priorities are identical.
Neighborhood Formation/Selection	Correlation weight threshold (CWT)	m neighbors for which $\text{sim}(a,c) \geq threshold_{cw}$
	Maximum number of neighbors (MNN)	$m = M$ neighbors with $\max(\text{sim}(a,c))$
Combining Ratings for Prediction	Simple arithmetic mean	$U^a(s) = \dfrac{\sum_{d=1}^{m} U^d(s)}{m}$
	Weighted mean	$U^a(s) = \dfrac{\sum_{d=1}^{m} U^d(s) \cdot \text{sim}(a,d)}{\sum_{d=1}^{m} \text{sim}(a,d)}$
	Deviation-from-mean	$U^a(s) = \overline{U^a} + \dfrac{\sum_{d=1}^{m} [(U^d(s) - \overline{U^d}) \cdot \text{sim}(a,d)]}{\sum_{d=1}^{m} \text{sim}(a,d)}$

one for each criterion g_i according to the way the users in C have previously evaluated items on each criterion g_i. The similarity of the active user a to a user $c \in C$ for criterion g_i is denoted as $\text{sim}^{g_i}(a,c)$ and takes into consideration the y commonly co-rated items of the active users a and c. The n predictions $g_i^a(s)$ ($i = 1, \ldots, n$) are then used to compute the prediction of the total utility of the target item s, according to the formula:

$$U^a(s) = \sum_{i=1}^{n} w_i^a g_i^a(s). \tag{5.4}$$

The various design options for the PG algorithm are similar to the ones for the PW one (see Table 5.1), whereas instead of w_i^a and w_i^c, $g_i^a(s)$ and $g_i^c(s)$ are used respectively. Again, a detailed description may be found in.[11]

5.3.1.3. Similarity per partial utility (PU) algorithm

This algorithm calculates the prediction of the total utility $U^a(s)$ of a target item $s \in S$, by predicting separately each partial utility $u_i^a(s)$, and then synthesizing these predictions into a total utility value. The predictions are based on the similarity between the partial utilities of the active user with the partial utilities of the rest of the users, upon each one of the n criteria. More specifically, the algorithm calculates the n predictions of the partial utilities $u_i^a(s)$ of the target item $s \in S$ ($i = 1, \ldots, n$), and then sums them altogether to produce the total utility $U^a(s)$. Again, n neighborhoods $D_i \subseteq C$ are created, one for each criterion g_i according to the partial utilities u_i users in C have provided for g_i. The similarity of the active user a to a user c for criterion g_i which is denoted as $\text{sim}^{u_i}(a,c)$, takes again into consideration the y commonly corated items of the active users a and each user $c \in C$. The n predictions $u_i^a(s)$ ($i = 1, \ldots, n$) are then used to compute the prediction of the total utility of target item s, according to the formula:

$$U^a(s) = \sum_{i=1}^{n} u_i^a(s). \tag{5.5}$$

The various design options for the PU algorithm are similar to the ones for that of the PW (see Table 5.1), whereas instead of w_i^a and w_i^c, $u_i^a(s)$ and $u_i^c(s)$ are used respectively. Again, a detailed description may be found in.[11]

5.3.2. Nonpersonalized algorithms

Apart from the three neighborhoodbased algorithms presented above, five non-personalized algorithms are also being considered, in order to serve as comparison measures throughout evaluation experiments. In particular, the following nonpersonalized algorithms are examined:

- The *Random* algorithm: it randomly produces a prediction of $U^a(s)$, independently from what evaluations other users have provided in the past.
- The *Random Exist* algorithm: it randomly selects one of the utilities $U^c(s)$ that a previous user $c \in C$ has given to item s, and presented this as the predicted value.
- The *Arithmetic Mean (AriMean)* algorithm: it calculates a prediction as the arithmetic mean of all $U^c(s)$ that all other users $c \in C$ have provided, independently of how similar they are to the active user.
- The *Geometrical Mean (GeoMean)* algorithm: it calculates a prediction as the geometrical mean of all $U^c(s)$, independently of how similar they are to the active user, according to $U^a(s) = \sqrt[c_{max}]{U^1(s) \times \cdots \times U^{c_{max}}(s)}$.
- The *Deviation-from-Mean (Dev-from-Mean)* algorithm: it calculates a prediction as a deviation-from-mean average over all $U^c(s)$. This algorithm actually aims to predict for the active user, what the average deviation from the mean of his previous evaluations will be, based on the other users' evaluations. It is recommended by[15] as a very efficient nonpersonalized algorithm (although it introduces some personalization factor, since it bases the prediction upon the mean value of the active user's evaluations). The formula for calculating $U^a(s)$ from the utilities of the c_{max} other users is the following:

$$U^a(s) = \overline{U^a} + \frac{\sum_{c=1}^{c_{max}} \left(U^c(s) - \overline{U^c}\right)}{c_{max}}. \tag{5.6}$$

In Eq. (5.6), $\overline{U^a}$ is the mean value of the evaluations that user a has provided in other items, and $\overline{U^c}$ the mean value of other evaluations that user c has provided.

The next section will introduce the particular application context for which the proposed algorithms have been implemented and experimentally tested.

5.4. Case Study and Experimental Analysis

The application of Internet technologies to online transactions has lead to the amazing growth of Internet-based e-markets. With the advent of these e-markets, numerous opportunities for online business participants (sellers, buyers etc.) have opened up. E-markets are operating in different business sectors, and are offering a variety of services that facilitate product and information exchange, as well as support the all-in process of transactions from initial contracts and negotiation to settlement.[34] This leads to a large amount of complex information that can become overwhelming for a typical Internet user. From the potential customer's perspective, tasks such as searching, locating, comparing and selecting appropriate e-markets can be difficult and time-consuming. Such obstacles may be partially overcome by the development of appropriate e-market recommender systems that will help users to easily locate e-markets according to their specific needs and preferences.

Focusing on the particular business sector of agriculture, we aim to deploy an online observatory of e-markets with agricultural products. An initial prototype of this observatory, termed as the "eMaM: e-Market Metadata Repository" (http:// e-services.aua.gr/eMaM.htm), contains a collection of about 200 e-market descriptions and allows for searching or browsing based on e-market characteristics.[35] It is our aim to enhance the services provided by eMaM, by adding an e-market recommendation service that will be based on multiattribute collaborative filtering.

In this context, the members of the eMaM user community are expected to be evaluating their experience from using an e-market. Evaluations will be collected using well-accepted and validated evaluation instruments for e-commerce services. In the experiment to follow, these are collected upon the e-market quality dimensions of eTailQ[36] but other options are also under investigation (e.g. WebQual[37]). Thus, the e-market recommender of eMaM will be taking as input multi-criteria evaluations from users, and will try to predict the e-markets that some particular user might like. Since the studied evaluation dimensions are the ones of eTailQ[36] the criteria set corresponds to the four evaluation dimensions that eTailQ uses to assess the quality of an e-market. All criteria take values from a 7-point scale $\{1,\ldots,7\}$, where "1" is the lower value of the criterion, and 7 the higher one. The developers of eTailQ claim that these criteria are independent, exhaustive, and non-redundant. Other criteria sets may also be selected, without affecting the design of the algorithms.

5.4.1. *Experimental setting*

The goal of the experimental testing has been twofold: first, to evaluate which of the three proposed algorithms is more appropriate for the particular eMaM application context; second, to examine the appropriate parameterization of the proposed algorithms, by exploring the various design options. For each stage of Sec. 5.3, the considered design options led to a number of algorithm variations (selected from the options in Table 5.1).

For this experiment, we have considered all three options of Similarity Calculation (*Stage A*), that is Euclidian, Vector/Cosine, and Pearson. We have not considered some particular method for Feature Weighting (*Stage B*), thus this factor was set equal to "1". Both methods for Neighborhood Formation/Selection (*Stage C*) have been considered, that is *correlation weight threshold* or CWT and *maximum number of neighbors* or MNN. Finally, all three options for Combining Ratings for Prediction (*Stage D*) have been examined. This led to 3*1*2*3 = 18 variations of each one of the three proposed algorithms. To fine-tune the algorithms and explore their appropriate parameterization, we further varied the parameter value of the Neighborhood Formation/Selection stage. For CWT, values varied between "0" and "1" (leading to 21 variations). For MNN, values varied between "1" and "20" (leading to 20 variations). The overall number of variations considered have been $(21 * 18 + 20 * 18)/2 = 369$ (from which, 189 using CWT and 180 MNN). To facilitate the comparison of the results of the different algorithm variations, we developed a simulator of multiattribute utility collaborative filtering algorithms.[38] This software tool allowed us to parameterize, execute and evaluate all considered variations of the proposed algorithms.

Since there are no multi-criteria data sets publicly available (as they exist in the case of single-criterion ratings, e.g. the MovieLens, EachMovie and Jester data sets[10,39]), we used the online simulator to produce a data set similar to the one that the eMaM environment will have when the recommendation system is put into operation. Our aim is to deploy eMaM in order to support a community of users interested in e-markets with agricultural products. Potential users for eMaM will be mainly attracted from a larger community of people interested in the application of new technologies in the agricultural sector: the community of the European Federation of Information Technology Applications in Agriculture (EFITA). For instance, the EFITA mailing list consists of more than 2.000 people around Europe. Based on anecdotal data from an informal survey that has been carried out during previous EFITA conferences, it has been possible to estimate that about 10% of the EFITA community members are expected to register as users of eMaM. It would be sensible to estimate that 10% of the EFITA mailing list members (that is about 200 users) will register and use the eMaM environment. Upon registration, each user will be asked to evaluate three examples of existing agricultural e-markets using the criteria of eTailQ, in order for an initial pool of evaluations to be created. The eMaM users can also provide additional evaluations of agricultural e-markets as they continue using the system. Based on the above estimation, we decided to examine the following scenario of the eMaM operation: about 200 users will be registered to the eMaM environment, which contain about 200 items, and (since each user will be asked to evaluate three e-markets upon the four criteria of eTailQ), eMaM is expected to have at least 600 evaluations. Therefore, the simulator has been used to produce a data set of 200 users, 200 items, and 750 evaluations, from which 600 could serve as a training set for the algorithms and the remaining 150 as a testing set. In this way, the training set had similar characteristics with the ones of the above scenario.

The evaluations have been processed with the simulation environment, and have been split into a training and into a testing component (using a 80–20% split). The performance of each algorithm variation has been measured as follows. For each evaluation in the testing component, the user that had provided this evaluation was considered as the active user, and the evaluated e-market as the target item. Then, the algorithm tried to predict the total utility that the target item would have for the active user, based on the information in the training component.

For our experimental analysis, two particular performance evaluation metrics have been used (similarly to the analysis of the single-criterion collaborative filtering algorithms of[15]). Other metrics for recommender systems evaluation are discussed in.[10] The metrics used have been the following:

- *Accuracy*: To measure the predictive accuracy of the multicriteria algorithms, we calculated the mean-absolute error (MAE). MAE is the most frequently used metric when evaluating recommender systems. Herlocker et al.[10] have demonstrated that since it is strongly correlated with many other proposed metrics for recommender systems, it can be preferred as easier to measure, having also well understood significance measures.
- *Coverage*: To measure the coverage of the multicriteria algorithms, we

calculated the items for which an algorithm could produce a recommendation, as a percentage of the total number of items. Previous research[10] recommends the measurement of coverage in combination with accuracy.

The simulator compared the predicted utility with the actual one, and calculated the MAE from all evaluations in the testing set. Furthermore, it calculated coverage as the percentage of e-markets in the testing component for which the algorithm could calculate a prediction based on the data in the training component. Additionally, the time required for a prediction to be calculated has also been recorded.

5.4.2. Results

Comparison results of the algorithm variations are presented in Tables 5.2 and 5.3, as well as Figs. 5.1–5.4. More specifically, Table 5.2

presents the Accuracy of the Accuracy of the various algorithm variations, whereas Figs. 5.3 and 5.4 present their Coverage. In all figures, the Simple algorithms appear using a cross +, the PW variations using a circle o, the PG variations using a triangle Δ, and the PU variations using a rectangle □.

In Figure 5.1, the variations where the Correlation Weighting Threshold (CWT) option for neighborhood formation has been chosen are presented (where the threshold takes values from 0 to 1, with step 0.05). It can be noted that the PG variations are generally producing more accurate predictions in terms of MAE. All variations perform better than the Random one, but the rest of the Simple variations performs similarly (or sometimes even better) to some personalized ones. Nevertheless, the difference in MAE is rather small (between 0.05 and 0.45). This picture changes when the Max

Table 5.2. MAE for each nonpersonalized algorithm.

Variation	Pure Random	Random Exists	Ari Mean	Geo Mean	Dev-from-Mean
MAE	2.06348	0.86143	0.71813	0.73591	0.75715

Table 5.3. Coverage for each nonpersonalized algorithm.

Variation	Pure Random	Random Exists	Ari Mean	Geo Mean	Dev-from-Mean
Coverage	100%	100%	100%	100%	94.59%

Neighbors Number (MNN) option is chosen for neighborhood formation.

From Figure 5.2 it can be noted that, when MNN is engaged, again some PG variations generally perform better than the rest of the algorithms. Moreover, the PW variations seem to be producing higher MAE than the rest of the algorithms, and in some cases even worse than the Random variations. When MNN is engaged, the difference in MAE between the variations can sometimes get very high. For example, it gets between 1.2 and 1.4 comparing the worst of the PW variations with the best performing variations.

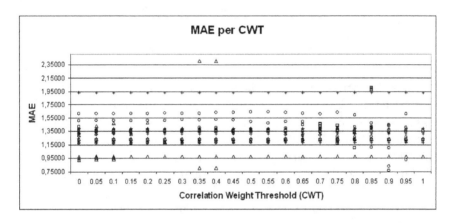

Fig. 5.1. MAE scatterplot for each CWT variation.

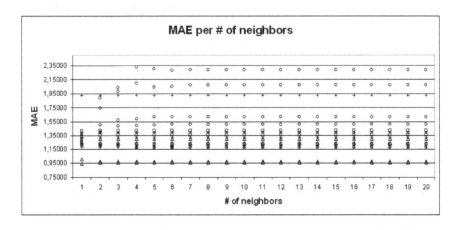

Fig. 5.2. MAE scatterplot for each MNN variation.

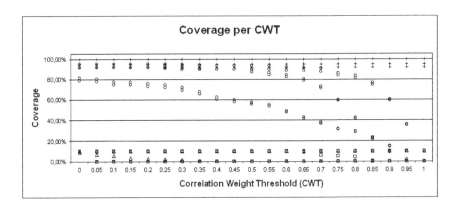

Fig. 5.3. Coverage scatterplot for each CWT variation.

Apart from the accuracy of the algorithms, it is important to examine their performance in terms of coverage. Indeed, Figure 5.3 demonstrates that all the PG and PU variations have very low coverage (>20%) for the examined data set. This means that they are not appropriate for operation under circumstances similar to the ones we empirically investigate. In this diagram it is also illustrated that the coverage of some PW variations decays as the CW threshold gets higher. On the other hand, there are also some PW variations that have high coverage (>80%), for most CW threshold values. As it has been expected, the Simple variations have the highest coverage values, since they do not have particular requirements for their training data e.g. the Random variation has 100% coverage since it always produces a prediction, and the AriMean variation has 95.33% coverage since it only requires the existence of some other users that have previously evaluated the target item.

In a similar manner, Figure 5.4 presents the coverage for all MNN variations. From this diagram, it can be noted that again all the PG and PU variations have very low coverage (around 10%) for the examined data set. It is also clear that the all PW variations have very high coverage (over 92%). The conclusion from the Figs. 5.3 and 5.4 is that for the particular data set, the PG and PU are not appropriate, although they have lower MAE in some occasions. From the above diagrams, it appears that the PW variations, with the MNN option chosen for neighborhood formation, seem more appropriate for the examined data set. To validate this observation, we compared all variations and identified the top-5 ones variations in terms of accuracy, which had also coverage equal or greater than 80%. Table 5.2

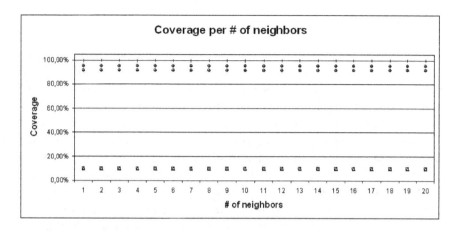

Fig. 5.4. Coverage scatterplot for each MNN variation.

Table 5.4. Top-5 algorithm variations according to MAE (with coverage > 80%).

Rank	VERSION	Neighb Method	Normalization	MAE	Coverage	Execution Time
1st	Pearson PW	MMN = 4	Simple Mean	1.164581	95.33%	<1 s
2nd	Pearson PW	MMN = 5	Simple Mean	1.164804	95.33%	1 s
3rd	Cosine PW	MMN = 4	Simple Mean	1.167469	95.33%	<1 s
4th	Euclidian PW	MMN = 4	Simple Mean	1.16778	95.33%	<1 s
5th	Euclidian PW	MMN = 5	Simple Mean	1.16871	95.33%	1 s

demonstrates that these are all PW variations, with the MNN option for the neighborhood formation, and illustrates how they perform in terms of accuracy, coverage and execution time.

From these results, it has been possible to identify which recommendation algorithm variation seems most appropriate for the particular application context. It has been demonstrated, that a PW algorithm variation that engages the Pearson metric for the calculation of similarity between user preferences, and the MNN method for the selection of neighborhood (with MNN=4), will provide a combination of high accuracy (prediction with MAE of about 1.165 on the eTailQ scale of 1 to 7) and high coverage (producing a prediction for about 95% of the e-markets) for the data set that is expected in the eMaM environment.

5.5. Discussion

In this paper, the proposed multicriteria recommendation algorithms are based on MAUT principles and use a linear additive value function for the representation of user preferences. This is a traditional decision making approach, widely applied and convenient to implement. On the other hand, assuming that the utility function is linear restricts the way user preferences are represented, since partial utilities are not always linear (e.g. they can be represented as sigmoid functions[40]). Therefore, we plan on exploring alternative utility function representations.[9,41] Another limitation of the MAUT approach is that it requires the user to fully express preferences upon criteria, as a set of importance weights. This can be addressed by the exploration of methods to elicit user preferences from past selections or that can calculate the similarity between user preferences, even when they are partially expressed.[24,32]

The major advantage of considering multiple criteria when producing a recommendation is the fact that users take more than one criteria into consideration for deciding whether an item is interesting/suitable for them. Furthermore, collaborative filtering may benefit from recommending items to users based on the items that users with similar preferences upon the multiple criteria have liked (instead of considering all users as candidate neighbors). Engaging multiple criteria may also allow for the exploration of alternative recommendation forms. For example, instead of recommending a user the items with the top-N total utility values, the items with the best combination of partial utility values upon specific criteria can be proposed (e.g. "the e-markets best scoring in the Reliability and the Website Design criteria").[33] The production of such recommendations would call for the use of more complex modeling methodologies, such as combinatorial/multiobjective optimization ones.

The proposed MAUT-based algorithms are neighborhood-based collaborative filtering ones. These algorithms have several benefits, such as their wide application and extensive testing, which allows for a better understanding of their expected behavior and performance. For instance, related studies have indicated that they produce rather precise and reliable results, even when compared to more sophisticated recommendation approaches.[39] On the other hand, they have well-known shortcomings, such as the fact that they do not perform well in sparse data sets and that they suffer from the "new user" and "new item" problems.[1,5] For this purpose, several improvements have been proposed in the literature, such as default voting and case amplification,[12] significance weighting of neighbors,[15] weighted-majority prediction,[42] as well as matrix conversion and instance-selection.[43] We have studied the extension of the presented algorithms in order to include these improvements in our proposed algorithms as well.[11] The next step is also to investigate if they may improve multiattribute collaborative filtering in the eMaM context. Other types of algorithms may also be explored, such as algorithms that are based on item-to-item correlations[13,44] or that are facing the recommendation problem as a multiobjective optimization problem.[1]

The number of criteria considered can greatly affect the performance of a multi-criteria algorithm. From other experiments with the proposed algorithms, where other sets of evaluation criteria have been also tested, it has been noted that the accuracy of the multi-criteria algorithms increased with the number of criteria. Furthermore, different design options may arise as more appropriate when different criteria sets are used. For example, in experiments where more than twenty criteria were used, the Euclidian variations seemed to be performing better than the other ones. On the other hand, as the number of criteria grew, the execution time of the algorithms had the tendency to grow significantly. This observation outlines the importance of carrying out a systematic evaluation and fine-tuning of several candidate algorithms, either on an existing data set with real evaluations or on a synthetic data set, before a multi-criteria recommender system is deployed in actual operation settings.

The evaluation metrics used in our experimental analysis (i.e. accuracy and coverage) are appropriate for the evaluation of recommender systems where prediction accuracy is important for the production of the recommendation. On the other hand, in several recommendation applications (such as top-N recommenders) the ranking accuracy is more important than the prediction accuracy (that is, the prediction of the correct item ordering is more important than exact utility value). Thus, we intend to extend our experimental analysis in order to examine the ranking accuracy of the algorithms, in usage scenarios where rankings of items are proposed to the users. Finally, the investigation of a combined metric that will synthesize accuracy, coverage and execution time in one formula, would make the selection of an algorithm that is appropriate for our application context much easier.

5.6. Conclusions

Careful testing and parameterization of a recommender system is required, before it is actually deployed in a real setting. Until today, very few multi-criteria recommender systems (e.g.[23]) have been systematically tested in the context of real-life applications. In this paper, we presented the experimental analysis of several design options for three proposed multi-attribute collaborative filtering algorithms for a particular application context, using a synthetic data set. The multi-criteria evaluations of the synthetic data set have been created using an appropriate simulation environment[38] and tried to be similar to the ones expected to be collected from actual users.

The results of this study provide useful insight about how a synthetic data set may be created and used in order to facilitate the study and selection of an appropriate recommendation algorithm, when actual multi-criteria evaluations from real users are not available. It has been highlighted that the performance of recommendation algorithms seems to be dependent on the application context, as they are reflected on the properties of the evaluations data set. Therefore, it is important to experimentally analyze various design choices for a multi-criteria recommender system, before its actual deployment in a real setting. On the other hand, experimental testing based on such a synthetic data set can only serve as for initial evaluation purposes: it may help a recommender system designer to decide about several design options regarding the recommendation algorithm to be used. As soon as evaluations from actual users are available, experimental analysis should be repeated to verify that the selected design options match the actual needs of the application context. A future research direction of particular interest would therefore be the exploration of adaptive recommender systems. Such a system will be able to dynamically select the appropriate recommendation algorithm or variation according to the current properties of the evaluations data set.

References

1. G. Adomavicius and A. Tuzhilin, Towards the next generation of recommender systems: a survey of the state-of-the-art and possible extensions, *IEEE Trans. Knowl. Data Engin.* **17**(6), 734–749, (2005).
2. B. N. Miller, J. A. Konstan, and J. Riedl, Pocketlens: toward a personal recommender system, *ACM Trans. Inform. Syst.* **22**(3), 437–476, (2004).
3. D. Goldberg, D. Nichols, B. M. Oki, and D. Terry, Using collaborative fil-

tering to weave an information tapestry, *Commun. ACM.* **35**(12), 61–70, (1992).
4. P. Resnick and H. R. Varian, Recommender systems, *Commun. ACM.* **40**(3), 56–58, (1997).
5. R. Burke, Hybrid recommender systems: survey and experiments, *User Model & User Adapt. Inter.* **12**, 331–370, (2002).
6. P. Perny and J. Zucker, Preference-based search and machine learning for collaborative filtering: the "film-conseil" movie recommender system, *Inform. Interact. Intell.* **1**(1), 9–48, (2001).
7. G. Tewari, J. Youll, and P. Maes, Personalized location-based brokering using an agent-based intermediary architecture, *Dec. Supp. Syst.* **34**, 127–137, (2002).
8. D. Liu and Y. Shih, Integrating ahp and data mining for product recommendation based on customer lifetime value, *Inform. Manag.* **42**, 387–400, (2005).
9. B. Price and P. R. Messinger. Optimal recommendation sets: covering uncertainty over user preferences. In *Proc. Inform. Ann. Meet. DenveAI Press*, (2005).
10. J. L. Herlocker, J. A. Konstan, L. G. Terveen, and J. T. Riedl, Evaluating collaborative filtering recommender systems, *ACM Trans. Inform. Syst.* **22** (1), 5–53, (2004).
11. N. Manouselis and C. Costopoulou, *Designing multiattribute utility algorithms for collaborative filtering algorithms*. (Technical Report, Informatics Laboratory, Agricultural University of Athens, TR 181, 2006).
12. J. S. Breese, D. Heckerman, and C. Kadie. Empirical analysis of predictive algorithms for collaborative filtering. In *Proc. 14th Conf. Uncertainty in Artificial Intelligence*, Madison, WI, USA. (July, 1998).
13. M. Deshpande and G. Karypis, Item-based top-n recommendation algorithms, *ACM Trans. Inform. Syst.* **22**(1), 143–177, (2004).
14. M. Papagelis and D. Plexousakis, Qualitative analysis of user-based and item-based prediction algorithms for recommendation agents, *Engin. Apps Art. Int.* **18**, 781–789, (2005).
15. J. Herlocker, J. Konstan, and J. Riedl, An empirical analysis of design choices in neighborhood-based collaborative filtering algorithms, *Inform. Retr.* **5**, 287–310, (2002).
16. R. L. Keeney, *Value-focused Thinking: A Path to Creative Decisionmaking.* (Harvard University Press, Cambridge, MA, 1992).
17. J. A. Konstan, Introduction to recommender systems: algorithms and evaluation, *ACM Trans. Inform. Syst.* **22**(1), 1–4, (2004).
18. B. Roy, *Multicriteria Methodology for Decision Aiding.* (Kluwer Academic Publishers, 1996).
19. P. Vincke, *Multicriteria Decision-Aid.* (J. Wiley, New York, 1992).
20. E. Jacquet-Lagreze and Y. Siskos, Preference disaggregation: 20 years of mcda experience, *Eur. J. Oper. Res.* **130**(233–245), (2001).
21. M. Stolze and M. Stroebel. Dealing with learning in ecommerce product navigation and decision support: the teaching salesman problem. In *Proc.*

2nd World Congr. Mass Custom. Person., Munich, Germany, (2003).
22. C. Schmitt, D. Dengler, and M. Bauer. The maut-machine: an adaptive recommender system. In Proc. ABIS Workshop Adaptivit{at und Benutzermodellierung in interaktiven Softwaresystemen, Hannover, Germany (October, 2002).
23. M. Montaner, B. Lopez, and J. L. de la Rosa. Evaluation of recommender systems through simulated users. In Proc. 6th Int. Conf. Enterprise Information Systems (ICEIS'04), pp. 303–308, Porto, Portugal, (2004).
24. V. Schickel-Zuber and B. Faltings. Hetereogeneous attribute utility model: a new approach for modelling user profiles for recommendation systems. In Proc. Workshop Knowl. Discov. Web, Chicago, Illinois, USA (August, 2005).
25. S. Guan, C. S. Ngoo, and F. Zhu, Handy broker: an intelligent product-brokering agent for m-commerce applications with user preference tracking, Electr. Commun. Res. Appl. **1**, 314–330, (2002).
26. W. P. Lee, Towards agent-based decision making in the electronic marketplace: interactive recommendation and automated negotiation, Exp. Syst. Appl. **27**, 665–679, (2004).
27. H. Nguyen and P. Haddawy. Diva: applying decision theory to collaborative filtering. In Proc. AAAI Workshop Recommend. Syst, Madison, WI, USA (July, 1998).
28. K. Yu, Z. Wen, X. Xu, and M. Ester. Feature weighting and instance selection for collaborative filtering. In Proc. 2nd Int. Workshop on Management of Information on the Web — Web Data and Text Mining (MIW'01), (2001).
29. D. W. Aha, D. Kibler, and M. K. Albert, Instance-based learning algorithms, Mach. Learn. **6**, 37–66, (1991).
30. T. M. Cover and P. E. Hart, Nearest neighbor pattern classification, IEEE Trans. Inf. Th. **13**(1), 21–27, (1967).
31. P. Resnick, N. Lacovou, M.Suchak, P. Bergstrom, and J. Riedl. Grouplens: an open architecture for collaborative filtering. In Proc. ACM CSCW, pp. 175–186, (1994).
32. V. Ha and P. Haddawy, Similarity of personal preferences: theoretical foundations and empirical analysis, Artif. Intell. **146**(2), 149–173, (2003).
33. G. Adomavicius, R. Sankaranarayanan, S. Sen, and A. Tuzhilin, Incorporating contextual information in recommender systems using a multidimensional approach, ACM Trans. Inform. Syst. **23**(1), 103–145, (2005).
34. Y. Bakos, The emerging role of electronic marketplaces on the internet, Commun. ACM. **41**(8), 35–42, (1998).
35. N. Manouselis and C. Costopoulou, Designing an internet-based directory service for e-markets, Inform. Serv. Use. **25**(2), 95–107, (2005).
36. M. Wolfinbarger and M. C. Gilly, etailq: dimensionalizing, measuring and predicting etail quality, J. Retailing. **79**, 183–198, (2003).
37. S. J. Barnes and R. Vidgen, An integrative approach to the assessment of e-commerce quality, J. Electr. Commun. Res. **3**(3), 114–127, (2002).
38. N. Manouselis and C. Costopoulou, Designing a web-based testing tool for multi-criteria recommender systems, Eng. Lett. Special Issue on Web Engineering. **13**(3), (2006).

39. L. Maritza, C. N. Gonzalez-Caro, J. J. Perez-Alcazar, J. C. Garcia-Diaz, and J. Delgado. A comparison of several predictive algorithms for collaborative filtering on multivalued ratings. In *Proc. 2004 ACM Symp. Applied Computing (SAC'04)*, Nicosia, Cyprus (March, 2004).
40. G. Carenini. User-specific decision-theoretic accuracy metrics for collaborative filtering. In *Proc. "Beyond Personalization" Workshop, Intelligent User Interfaces Conference (IUI'05)*, San Diego, California, USA (January, 2005).
41. J. Masthoff. Modeling the multiple people that are me. In eds. P. Brusilovsky, A. Corbett, and F. de Rosis, *Proc. User Modeling 2003, Lecture Notes in Artificial Intelligent*, pp. 258–262. Springer, Berlin, (2003).
42. J. Delgado and N. Ishii. Memory-based weighted-majority prediction for recommender systems. In *Proc. ACM-SIGIR'99, Recommend. Syst. Workshop*, UC Berkeley, USA (August, 1999).
43. C. Zeng, C. Xing, L. Zhou, and X. Zheng, Similarity measure and instance selection for collaborative filtering, *nt. J. Electr. Comm.* **8**(4), 115–129, (2004).
44. B. Sarwar, G. Karypis, J. Konstan, and J. Riedl. Analysis of recommendation algorithms for e-commerce. In *Proc. ACM EC'00*, Minneapolis, Minnesota, (2000).

BIOGRAPHIES

Nikos Manouselis received a diploma in electronics and computer engineering in 2000, a M.Sc. degree in operational research in 2002, and a M.Sc. degree in electronics and computer engineering in 2003, all from the Technical University of Crete (Greece). Since 2005, he is a Ph.D. candidate at the Informatics Laboratory of the Agricultural University of Athens (Greece).

Constantina Costopoulou received a B.Sc. degree in mathematics from the National and Kapodistrian University of Athens (Greece) in 1986, the M.Sc. degree in software engineering from Cranfield Institute of Technology (UK) in 1989, and the Ph.D. in electrical and computer engineering from the National Technical University of Athens (Greece) in 1992. Since 1995, she is an Assistant Professor at the Agricultural University of Athens (Greece).

Chapter 6

Efficient Collaborative Filtering in Content-Addressable Spaces

Shlomo Berkovsky*, Yaniv Eytani[†] and Larry Manevitz[‡]

Computer Science Department, University of Haifa
31905, Mount Carmel, Haifa, Israel
*slavax@cs.haifa.ac.il
[†]ieytani@cs.haifa.ac.il
[‡]manevitz@cs.haifa.ac.il

Collaborative Filtering (CF) is currently one of the most popular and most widely used personalization techniques. It generates personalized predictions based on the assumption that users with similar tastes prefer similar items. One of the major drawbacks of the CF from the computational point of view is its limited scalability since the computational effort required by the CF grows linearly both with the number of available users and items. This work proposes a novel efficient variant of the CF employed over a multidimensional content-addressable space. The proposed approach heuristically decreases the computational effort required by the CF algorithm by limiting the search process only to potentially similar users. Experimental results demonstrate that the proposed heuristic approach is capable of generating predictions with high levels of accuracy, while significantly improving the performance in comparison with the traditional implementations of the CF.

6.1. Introduction

In many circumstances, the quantity of available information grows rapidly and exceeds our cognitive processing capabilities. Thus, there is a pressing need for intelligent *personalization* systems providing services tailored to users' real needs and interests. Recommender Systems[1] are one of the commonly used approaches to address this problem. These systems assist a user in selecting a suitable item among a set of potentially selectable items by predicting the user's opinion on the items by applying statistical and knowledge discovery techniques.[2] Currently, Recommender Systems are used in a variety of application domains, e.g. movies,[3] jokes,[4] music[5] and others, and they exploit various recommendation techniques, such as Collaborative Filtering,[6] Content-Based Filtering,[7] Case-Based Reasoning,[8] and numerous hybrid techniques.[9]

Collaborative Filtering (CF) is probably the most familiar and one of the most widely-used techniques to generate predictions in Recommender Systems. It relies on the assumption that people who agreed in the past will also agree in the future.[10] The input for the CF algorithm is a matrix of users' ratings on a set of items, where each row represents the ratings provided by a single user and each column represents the ratings provided by different users on a single item. CF aggregates the ratings to recognize similarities between users and generates the prediction for an item by weighting the ratings of similar users on this item.

The CF algorithm is typically partitioned to three generic stages: (1) Similarity Computation: weighting all the users with respect to their similarity with the *active user* (i.e. the user, whose ratings are being predicted), (2) Neighborhood Formation: selecting the most similar users for the prediction generation, and (3) Prediction Generation: computing the prediction by weighting the ratings of the selected users.

One of the major drawbacks of the CF is its limited scalability. The stages of Similarity Computation and Neighborhood Formation require comparing the active users with all the other users over all the available ratings. Hence, the computational effort required by the CF grows linearly both with the number of users and the number of items in the ratings matrix. Thus, for a matrix containing ratings of M users on N items, the required computational effort is $O(MN)$. This poses a problem in real-life systems, where the predictions are generated using millions of ratings on thousands of items, e.g. in Web-based Recommender Systems. Previous studies, (e.g. Refs.,[11][12] and[4] and others) tackle the issue of reducing the computational effort required by the CF either by preprocessing of the ratings matrix or by distributing the heavy computational stages. Nonetheless it remains one of the most important issues in the CF research community.

In this work we develop a fast heuristic variant of the CF algorithm that decreases the computational effort required by the Similarity Computation and the Neighborhood Formation stages. The basic assumption of the proposed heuristic algorithm is that losing general completeness of the exhaustive search (1) has a minor negative effect on the accuracy of the predictions, but (2) significantly decreases the required computational effort. Thus it provides a scalable approach, applicable to real-life scenarios with a high number of users and items, such as in Web-based systems.

The proposed heuristic approach is based on a notion of content-addressable data management[13] that provides an adaptive topology for mapping of users' profiles to a multidimensional space. This mapping implicitly clusters similar users and limits the Similarity Computation and the Neighborhood Formation stages to a heuristic search among the users that are potentially highly similar to the active user.

Experimental evaluation of the proposed approach demonstrates both high efficiency and good accuracy of the proposed algorithm in comparison with the traditional (exhaustive) *K-Nearest Neighbors* (KNN) search of the Neighborhood Formation stage. The evaluation also demonstrates that the algorithm is highly scalable with the number of nearest neighbors to be retrieved.

The rest of the paper is organized as follows. Section 6.2 describes the CF personalization technique and surveys the studies focusing on the required computational effort reduction. Section 6.3 describes the CAN, a Peer-to-Peer content-addressable platform for decentralized data management. Section 6.4 describes the decentralized storage of users' profiles over the CAN platform and elaborates on the proposed heuristic variant of the CF over CAN. Section 6.5 presents and analyzes the experimental results. Finally, Sec. 6.6 lists our conclusions and presents some open questions for future research.

6.2. Collaborative Filtering

Collaborative Filtering (CF) is probably one of the most familiar and widely-used recommendation techniques. An input for the CF is the so-called *ratings matrix*, where each user is represented by a set of explicit ratings given on various items, and each item is represented by a set of ratings given by the users.

CF requires a similarity metric between users to be explicitly defined. The state-of-the-art CF systems exploit three similarity metrics: Cosine Similarity,[3] Mean Squared Difference (MSD),[14] and Pearson correlation.[2] This work focuses on the MSD, computing the degree of similarity between users x and y by:

$$\text{sim}_{x,y} = \frac{\sum_{i=1}^{|x \cap y|} (R_{x,i} - R_{y,i})^2}{|x \cap y|} \qquad (6.1)$$

where $|x \cap y|$ denotes the number of items rated by both users (typically, above some minimal threshold), and $R_{x,i}$ denotes the rating of user x on item i. In some sense, $\text{sim}_{x,y}$ can be considered as the *dissimilarity* of the users, as the lower the result of the MSD computation, the greater is the real similarity between the users.

Prediction $P_{a,j}$ for the rating of the user a on item j is computed as a weighted average of the ratings of his/her K most similar users, i.e. K *nearest neighbors*, by:

$$P_{a,j} = R'_a + \frac{\sum_{k=1}^{K} (R_{k,j} - R'_k) \cdot \text{sim}_{a,k}}{\sum_{k=1}^{K} |\text{sim}_{a,k}|} \quad (6.2)$$

where $R_{x,y}$ denotes the rating of user x on item y, R'_z denotes the average rating of user z, and $\text{sim}_{v,u}$ denotes the level of similarity between users v and u.

The Similarity Computation stage of the CF requires comparing the active user with every other user in the system. For a ratings matrix storing the ratings of M users on N items, the computational complexity of the Similarity Computation stage is $O(MN)$. This indicates poor scalability of the Similarity Computation stage, as the complexity grows linearly with both the number of users and the number of items in the matrix.

6.2.1. *Reducing the computational effort required by the CF*

Many prior works have dealt with decreasing the computational, effort required by the CF. In general, it is achieved either by preprocessing the ratings matrix, or by distributing the computationally intensive stages of the CF among multiple machines.

Various preprocessing techniques for decreasing the computational effort required by the CF (e.g. correlation coefficients, vector-based similarity, and statistical Bayesian methods) are discussed and analyzed in Ref.[11] Another technique, exploiting preclustering of the ratings matrix, is discussed in Ref.[4] There, principal component analysis is used to identify two *discriminative* dimensions of the ratings matrix and all the vectors are projected onto the resulting plane. This inherently partitions the users to clusters or neighborhoods, which are further used to generate the predictions. In Ref.,[12] the authors use a tree-like data structure and apply a *divide-and-conquer* approach using an iterative K-means clustering to group the users. This leads to smaller and more homogeneous clustering of users for the following Predictions Generation stage.

An alternative approach is to distribute the computational effort required by the CF among the users, such that every user independently computes its similarity with the active user. This approach was initially proposed in Ref.[15] and elaborated in Ref.[16] The latter also developed a detailed taxonomy of the CF distribution approaches and presented implementation frameworks for different application domains. The PocketLens project[17] compared five decentralized distributed architectures for the CF. These comparisons showed that the performance of the decentralized mechanism is similar to the performance of the centralized CF while providing increased robustness and security.

Further improvements to the decentralized CF were discussed in Ref.,[18] which proposes the exploitation of Peer-to-Peer platform for a decentralized management of users' profiles. However, this approach approximates the set of the most similar users identified by the Neighborhood Formation stage of the CF, and as a result, the accuracy of the generated predictions is reduced.

This paper is loosely based on the ideas of CAN,[13] a content-addressable Peer-to-Peer platform. We implement a fast heuristic variant of the CF, using a CAN-like multidimensional space for maintaining a connected structure of users. This allows to significantly decrease the computational effort required by the Similarity Computation and Neighborhood Formation stages by limiting the search process to a search among potentially similar users located in close vicinity to the active user.

6.3. Content-Addressable Data Management

This section presents the general architecture of CAN,[13] a scalable decentralized data management platform. In CAN, the users are represented in a one-to-one manner by the nodes of a virtual N-dimensional coordinate space such that the location of the user's node is denoted by a vector $(v_1, v_2, \ldots, v_N)$, where v_i represents the numeric coordinate of the node within a dimension number i. In addition to the node, each user continuously manages an N-dimensional subspace, called a *zone*. For example, consider a two-dimensional space partitioned to three zones, managed by users A, B, and C [Fig. 6.1 (left)]. Note that the figure shows only the zones managed by the users, whereas the nodes themselves are not shown.

In CAN space, two nodes (and also zones) are called *neighbors* if their coordinate spans overlap along $N - 1$ dimensions and adjoin along one

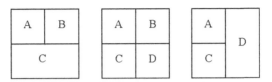

Fig. 6.1. Example of a two-dimensional CAN space.

dimension. For example, consider the neighbor zones A and C in Fig. 6.1 (left), whose coordinates partially overlap across the horizontal dimension and adjoin along the vertical. To maintain connectivity in CAN space, each node stores a data structure representing a list of pointers to a set of other nodes, managing the neighbor zones. For example, node A stores the pointers to the nodes managing zones B and C (as, respectively, horizontal and vertical neighbors) in its list of pointers.

Routing of messages in CAN space is based on the Plaxton routing algorithm.[19] This routing iteratively forwards the messages to the nodes that are closer to the target node than the current node using a greedy forwarding. The metric for evaluating the distance between two nodes in the address space is the L_1 metric, i.e. the Manhattan Distance. This metric was chosen due to the fact that CAN space inherently supports it, as every node stores a list of pointers to the nodes, managing the neighbor zones. For example, the distance between the nodes $(1,2,3)$ and $(6,5,4)$ in three-dimensional CAN space is $(6-1)+(5-2)+(4-3)=9$. Thus, in N-dimensional CAN space a message is routed between an arbitrary pair of nodes in $O(N)$ routing steps.

In addition, CAN provides a connectivity maintenance algorithm, stable to sporadic joints and departures of new users. When a new user is inserted, it is assigned its own node and the respective zone. This is done by splitting a zone (determined by the content provided by the recently inserted user) of one of the existing neighbors according to the following steps: (1) the new user identifies an existing network node, (2) the new user is routed to the target zone that will be split, and (3) the target zone is split and the neighbors of the new zone are updated to maintain connectivity and facilitate future routings. As a result, only a subset of immediate neighbor zones of the zone that was split is actually affected by the insertion of a new node.

The issue of splitting the target zone (i.e. how to split the existing zone, where the contents of the recently inserted node are mapped?) is one of the important issues affecting the performance of CAN. A number of splitting policies are proposed, analyzed and compared in Ref.[13] The simplest policy for the zones splitting is so-called *ordered* splitting. According to this policy, the number of dimension, across which a zone is split, iteratively increases from 1 to N.

For example, consider user D joining CAN two-dimensional space [Fig. 6.1 (middle)]. Assuming that the content provided by user D should be located in the right part of the zone managed by node C and this is the zone that will be split, user D is routed to C using the Plaxton routing, and zone C is split across the horizontal dimension (assuming that the previous split of zone C, and also the following split of both zones C and D will be performed across the vertical dimension). Finally, the recently inserted node, managing the zone D notifies its neighbors (i.e. the users "managing zones B and C) about the insertion of a new node, and also their neighbors" pointer tables are updated. Note that in this case, only the zone managed by user C, which was split and a subset of its neighbor zones (actually, only one zone managed by user B), are affected by the insertion of a new user D, whereas other zones are not affected.

Disconnections of the users are handled in a similar manner. The disconnecting user identifies one of the neighbor nodes that will takeover managing its zone, and updates other neighbor zones about the departure and the management takeover. For example, consider the user managing zone B disconnecting from CAN space [Fig. 6.1 (right)]. As a result of the disconnection, the user managing zone D takes over the management of the zone previously managed by user B.

Nonetheless, it should be mentioned that the dimensionality of the above CAN space can be barely extended. Such extension requires remapping of the existing N-dimensional nodes to a new $(N+1)$-dimensional space. This is an expensive procedure in a decentralized P2P environment, which can also involve multiple interactions with the users. Thus, in this work we assume that the dimension of CAN space is fixed.

Thus, CAN provides a decentralized platform, supporting (1) dynamic space partitioning and zones allocation, (2) efficient routing algorithm, and (3) connectivity maintenance algorithm over virtual N-dimensional coordinate space. Note that the distributed structure of CAN is not robust against sudden departures of users, as fault-tolerance is not one of the main goals of the platform. However, CAN facilitates a decentralized self-manageable platform for content-addressable data management in a distributed environment.

6.4. CF over Content-Addressable Space

This work proposes an efficient heuristic variant of the CF algorithm. It uses a content-addressable architecture for the purposes of optimizing traditional exhaustive K-Nearest Neighbors (KNN) search to a search among potentially similar users only. Although our algorithm is a heuristic one by nature, experimental results demonstrate that it facilitates efficient search process without hampering the accuracy of the generated predictions.

6.4.1. *Mapping user profiles to content-addressable space*

The input for the CF algorithm is a matrix of users' ratings on items, where each row (ratings vector) represents the ratings of a single user and each column represents the ratings on a single item. The total number of items (N) defines an N-dimensional space, where the coordinates range in each dimension corresponds to the range of ratings on the respective item.

To handle the ratings matrix in a content-addressable manner, we map it to a CAN-like multidimensional space. Each rating is projected using a uniform *injective* (*one-to-one*) *mapping* onto the appropriate dimension, such that the whole vector of length N is mapped to a single point in an N-dimensional space. For example, consider a system storing the ratings of users on three different items. In such a system, the evolving CAN-like space will be a three-dimensional cube, where the range of coordinates within every dimension corresponds to the range of possible ratings on the respective item.

As already mentioned, each user is represented in a CAN-like space by a single node whose location corresponds to the set of user's ratings and by the respective zone (storing a list of immediate neighbor zones). For example, consider a user U that rated all three items in the above three-dimensional cube: item i_1 was rated as r_1, item i_2 as r_2, and i_3 as r_3. The user will be mapped to a location (r_1, r_2, r_3) of the space and will have exactly two neighbors in each dimension. For example, in the dimension corresponding to item i_1, the user U will have two neighbors, $N_1 = (r_1 - x, r_2, r_3)$ and $N_2 = (r_1 + y, r_2, r_3)$, such that both N_1 and N_2 rated i_2 as r_2 and i_3 as r_3, N_1 rated i_1 below r_1, and N_2 rated it above r_1, and there is no other user that rated i_1 as r', where $r_1 - x < r' < r_1$ or $r_1 < r' < r_1 + y$. Similarly, user U will have two neighbors in the dimension corresponding to item i_2 and to item i_3. If there is no user that provided the required combination of ratings on the available items, CAN space will maintain connectivity by connecting user U to a further node, which will serve as its immediate neighbor (i.e. both zones will keep mutual pointers to the relevant neighbor zone).

Note that in the evolving CAN space, the users (through their ratings vectors) can be dynamically inserted and removed not only during the initialization, but also during the life cycle of the system. This is explained by the observation that the above connectivity maintenance algorithm guarantees that the structure remains connected regardless of the sudden joints and disconnections of the nodes. Nevertheless, CAN spaces can barely manage insertions of new items, as the dimension of the space should remain fixed. Thus, the proposed heuristic search (that will be discussed in the following subsection) is applicable only over a stable matrix of users' ratings, where no new items are inserted.

Deciding on the zones split policy affects the evolving structure of the ratings vectors. In our implementation, we used the above mentioned ordered splitting policy. This policy may be suboptimal in terms of the number of neighbor zones, resulting in a less efficient algorithm, i.e. more comparisons or retrieving less similar neighbors. However, our experiments demonstrate that even this simple policy considerably increases the efficiency of the proposed K-Nearest Neighbors (KNN) search, in comparison with the traditional exhaustive search. Evaluating other splitting policies is beyond the scope of this work.

In addition to the guaranteed connectivity, content-addressable space *inherently clusters* similar users, such that the distance between two similar users (in our case, according to the MSD similarity metric) is lower than the distance between two arbitrary users. This is achieved due to the use of an injective mapping of the ratings vector to the multidimensional CAN-like space, which preserves the users' similarity while mapping the ratings vectors to the numeric coordinates in the space. The following subsection shows a use of the above inherent clustering property for the purposes of developing fast heuristic variant of the KNN search.

6.4.2. Heuristic nearest-neighbors search

The Neighborhood Formation stage of the CF over the evolving N-dimensional space can be schematically described as a heuristically expanding breadth-first search. The algorithm for retrieving K-Nearest Neighbors of a user x is briefly explained by the following pseudo-code. The code uses two lists of size K: (1) CANDIDATES — list of candidates for being one of the K-nearest neighbors, and (2) NEIGHBORS — list of real K-Nearest Neighbors. In principle, the algorithm needs the CANDIDATES list only, as the NEIGHBORS list only increases during the execution of the algorithm until it reaches its maximal length and contains the real K-Nearest Neighbors. For the sake of clarity, we show an algorithm that uses two lists instead of only one.

K_Nearest_Neighbors (*user x*)
(1) let NEIGHBORS and CANDIDATES be empty lists, each of size K
(2) let Z be the zone, to where x would be mapped in the CAN space
(3) foreach u$\in$ (Z$\cup$ neighbors(Z))
(4)   compute distance(x,u)
(5)   insert u into CANDIDATES, s.t. CANDIDATES is sorted
       according to the values of distances(x,u)
(6) for i=1 to K
(7)   choose v from CANDIDATES, s.t. distance(x,v) is smallest
(8)   for each w$\in$ neighbors(v) s.t. distance(x,w) is unknown
(9)   compute distance(x,w)
(10)   insert w into CANDIDATES, s.t. it remains sorted
           according to the values of distances(x,v)
(11)   move v from CANDIDATES to NEIGHBORS
(12) return NEIGHBORS

Initially, the algorithm pretends to map the active user x to its location in the N-dimensional space (step 2). Next, the algorithm identifies the zone x is mapped to, and its neighbors, i.e. users managing the neighbor zones (step 3). For each of these zones, the degree of similarity, i.e. the distance between x and the relevant user, is computed (step 4). Then, the neighbor users are inserted into the CANDIDATES list such that the whole list of candidates users is sorted according to the distances of the users from the active user x (steps 4 and 5).

Afterwards, the algorithm iteratively performs the following operations:

- Selects v, the nearest neighbor stored in the CANDIDATES list (step 7).
- Identifies the neighbors of v that are not in the CANDIDATES list yet, computes their distances from x, and inserts them into the CANDIDATES, while keeping the list sorted (steps 8–10).
- Removes v from the CANDIDATES list and inserts it into the NEIGHBORS list.

Finally, the algorithm returns the resulting NEIGHBORS list (step 12).

Consider an example execution of the KNN search as illustrated in Fig. 6.2. The initial structure of two-dimensional space is depicted in Fig. 6.2(a). Nine users, named from a to i, are inserted into the space and manage the respective zones. Note that also this figure shows only the zones managed by the users, whereas the nodes representing the users are not shown. Assume that the active user is mapped to the zone managed by user e.

Thus, e and its neighbors, i.e. users managing zones c, d, f and i, are the first candidates for being the nearest neighbors and they are inserted into the CANDIDATES list. Assume that the user managing zone e is the closest one. It is moved from the CANDIDATES list to the NEIGHBORS list [Fig. 6.2(a)]. Since all the neighbors of e are already known, the next closest neighbor is chosen among its neighbors. Assume that the next closest neighbor is the user managing zone f. It is moved from the CANDIDATES list to the NEIGHBORS list, and its only new neighbor, the user managing zone g, is inserted into the CANDIDATES list [Fig. 6.2(b)]. The next closest neighbor is the user managing zone c, inserting the user managing zone b into the CANDIDATES list [Fig. 6.2(c)]. Assume that the next closest neighbor is the user managing zone g (not an immediate neighbor of e). As a result, the user managing zone h is inserted into the CANDIDATES list [Fig. 6.2(d)]. This process is repeated until the NEIGHBORS list contains K-Nearest Neigh-

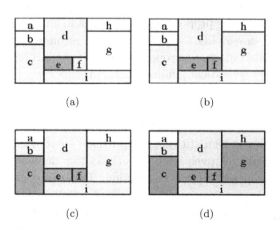

Fig. 6.2. Stages of the KNN search over two-dimensional CAN space (zones managed by users from the CANDIDATES are indicated with a light tone and from the NEIGHBORS — in a dark gray tone).

bors.

The proposed algorithm reduces the computational effort required by the Similarity Computation and the Neighborhood Formation stages, in comparison with the traditional CF algorithm, where an active user is compared with all the available users. Conversely, the proposed heuristic algorithm compares the active users with potentially similar users only, located in close vicinity to the active user.

Since every user in the N-dimensional space continuously maintains an updated list of its immediate neighbors, any neighbor of a given user is accessed through a single network hop. This is true regardless of the physical (geographical) and logical (similarity) distances between the neighbors. Thus, the algorithm will also work in sparse spaces, where the distance between neighbors in the underlying network might be very high.

6.4.3. *Heuristic completions of user profiles*

In the former sections, we assumed that the user's ratings were represented as a *complete vector*, i.e. explicit ratings on all the items are available. Thus the mapping of the user's ratings vectors to the underlying content-addressable space is straight-forward. However, this assumption is unachievable in most real-life applications and scenarios, where an average user rates only a portion of the available items. This raises a need for developing a mapping mechanism capable of mapping *incomplete vectors*, where a subset of the ratings is missing, to the content-addressable space.

In this subsection we propose three mappings to handle this task. However, instead of developing a new mapping of incomplete vectors to the content-addressable space, we propose to convert the incomplete vectors to complete ones by heuristically filling-in the missing ratings in the incomplete vectors.[20] Thus, the proposed completion heuristics are designed to reuse the above injective mapping of complete vectors, while employing it on the modified vectors with heuristically filled-in ratings.

As the completion heuristics are not the main focus of the current work, we suffice with three relatively simple heuristics that demonstrate the applicability of the proposed vectors' completion. The heuristics are as follows:

- *User-average* — The missing rating on an item in the user's vector is substituted with the average of the real ratings, explicitly provided by this user.
- *Item-average* — The missing rating on an item in the user's vector is

substituted with the average of the real ratings, explicitly provided by the other users on this item.
- *Conditional* — Integrates both the user-average and the item-average heuristics and decides on a run-time regarding the specific completion heuristic to be used according to a certain predefined condition.

Clearly, the *user-average* heuristic can be considered as an accurate personalized completion heuristic, as the missing ratings are substituted with a value, produced by the real ratings of the given user. Thus, it reflects the real preferences and tendencies of the user, such as over- or under-rating of items, natural intensity of expressions and so forth. Conversely, the *item-average* heuristic can be considered as the most accurate non-personalized completion heuristic, as the missing ratings are substituted with a value, produced by numerous real ratings on the given item. As such, it reflects a general (and relatively reliable) opinion of many other users on the item.

We conjecture that the *user-average* heuristic is preferable when the knowledge about the user's preferences is reliable, i.e. the number of ratings explicitly provided by the user is relatively high. On the other hand, when the number of user's explicit ratings is low, the *item-average* heuristic will exploit other users' ratings for filling-in the missing rating and it should be preferred. Based on these considerations, we defined another *conditional* heuristic, which will autonomously decide which of the above completion heuristics should be exploited for filling-in the missing ratings of every user.

In summary, each of these heuristics allows the filling-in of the missing ratings, converting the incomplete vectors to the complete ones, and then mapping them to the content-addressable space using the abovementioned injective mapping mechanism.

6.5. Experimental Results

In the experimental part of our work we used the Jester dataset of jokes' ratings.[4] Jester is a Web-based jokes Recommender System, containing 4.1 millions of ratings (on a continuous scale from -10.00 to $+10.00$) of 73,421 users on 100 jokes. A significant portion of the users rated all the jokes, so the Jester dataset is relatively dense. Overall, approximately 56% of all the possible ratings in the matrix are present.

For the complete vectors experiments, we selected a subset of 14,192 users that rated all 100 jokes, producing a matrix, where every value corresponds to a real rating, explicitly provided by a user. The average rating of a single joke in the data is 0.807, and the overall standard deviation of the ratings in the matrix is 4.267. We implemented a centralized simulation of a 100-dimensional CAN space (note that the space dimension equals to the number of rated jokes in the ratings vectors) and inserted the above 14,192 users into the space. Insertions of the users into the space were done using the ordered splitting policy.

6.5.1. Scalability of the search

These experiments were designed to evaluate the scalability of the proposed heuristic variant of the KNN search. The efficiency of CAN-based KNN is measured by the number of comparisons performed during the Neighborhood Formation stage of the CF.

In this experiment we measured number of comparisons during the Neighborhood Formation stage. For this, we gradually increased the number of users inserted into the system from $M = 1000$ to $M = 14,000$. For each M, we computed the number of comparisons performed in the traditional exhaustive KNN search and in CAN-based heuristic variant of KNN. Both searches were aimed at retrieving $K = 5$ nearest neighbors. For each value of M, the experiments were repeated 1000 times for different active users. The experimental results are shown in Fig. 6.3. The horizontal axis stands for M, the number of users inserted into the system, and the vertical axis reflects the average number of comparisons during a single KNN search, for both exhaustive and heuristic searches.

As expected, the number of comparisons in CAN-based KNN is significantly lower than in traditional KNN and it grows at a logarithmic-like manner with the number of users. This is explained by the fact that in CAN-based KNN the active user is compared only with a subset of highly similar users (located in close vicinity in a content-addressable space), whereas in traditional KNN it is exhaustively compared with all the available users.

To achieve a better understanding of comparison-based scalability of the proposed approach, we computed the ratio between the number of comparisons in CAN-based KNN and the number of comparisons in the exhaustive KNN. This ratio was computed for different values of M and the results are shown in Fig. 6.4. It can be clearly seen that the ratio steadily decreases with M. This allows us to conclude that the proposed algorithm is applicable in large-scale systems with high number of users and items, e.g. on the Web.

The second experiment was designed to evaluate the scalability of CAN-

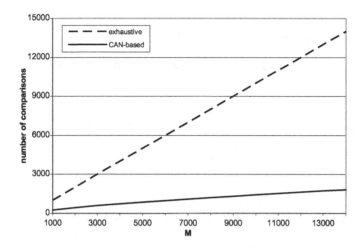

Fig. 6.3. Average number of comparisons versus the number of users inserted.

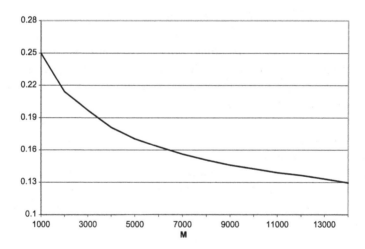

Fig. 6.4. Ratio between the number of comparisons versus the number of users inserted.

based KNN with the number of nearest neighbors (K) to be retrieved. We gradually increased the value of K from $K = 1$ to $K = 50$. For each value of K, we measured the number of comparisons needed to retrieve K nearest neighbors for $M = 1000$, 2000, 4000, 8000, and 14,000 users. For each value of M and K, the experiments were repeated 1,000 times for different active users. The number of comparisons as a function of K for the above values of M is shown in Fig. 6.5. The horizontal axis stands for K, the number of nearest neighbors to be retrieved, whereas the vertical reflects the average number of comparisons during the KNN search.

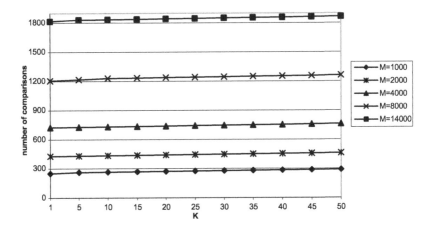

Fig. 6.5. Average number of comparisons versus the number of retrieved neighbors.

As can be clearly seen, the number of comparisons in CAN-based KNN remains roughly unchanged when K increases. This is explained by the observation that most of the KNN users are located in close vicinity to the active user (this characterizes a real-life naturally clustered data). Thus, the similar users are discovered in the early stages of the KNN search, while further expansions contribute very few new similar users.

Both experiments show good scalability of CAN-based KNN with K. This means, that practical Recommender Systems can use higher values of K, to form moderately larger and more reliable neighborhoods, and generate more accurate predictions with only a very minor computational overhead.

6.5.2. *Accuracy of the search*

The following experiments were designed to evaluate the accuracy of the results obtained by the proposed heuristic variant of KNN search. In the first experiment we compared the sets of users, i.e. the neighborhoods, retrieved by the traditional (exhaustive) KNN and by the CAN-based variant of KNN.

Let us denote by KNN_e the set of users retrieved by the traditional exhaustive KNN search and by KNN_h the set of users retrieved by the

CAN-based heuristic variant of KNN. Since the CAN-based KNN is a heuristic approach, a suboptimal structure of zones may lead to a situation, where $KNN_e \neq KNN_h$, i.e. the heuristic search retrieves only a subset of the real K nearest neighbors. As the collaborative predictions are generated by aggregating the ratings of similar users, identifying the set of most similar users is essential for generating accurate predictions.

To evaluate the accuracy of the proposed heuristic KNN search, we adapt the traditional Information Retrieval metric of precision.[21] In fact, the computed accuracy metric is not a classical precision, but rather precision@K, since the overall search procedure is limited to K most similar users only. However, this metric also provides some indication about the recall of the search, as it can be considered as the recall of the search for a limited number of the most similar users to be retrieved. For the sake of clarity, this metric is referred to in the paper as precision. The precision is computed by:

$$\text{precision} = \frac{|KNN_e \cap KNN_h|}{|KNN_e|} = \frac{|KNN_e \cap KNN_h|}{K}. \qquad (6.3)$$

The cardinality of the KNN_e set was $K = 10$, while the cardinality of the KNN_h set was gradually increased from $K' = 1$ to $K' = 100$. The precision was computed for $M = 1000, 2000, 4000, 8000$ and $14,000$ users inserted into the system. For each value of M and K', the experiments were repeated 1000 times for different active users. Figure 6.6 shows the precision as a function of K' for the above values of M. The horizontal axis stands for M, the number of users inserted into the system, whereas the vertical reflects the average precision of the heuristic KNN search.

As can be clearly seen, the curves behave similarly and the accuracy increases with K', such that for $K' > 50$, it is over 0.9 for all the given values of M. Previous experiments presented in previous subsection show that the algorithm is highly scalable with K. Thus, retrieving a larger set of users (i.e. higher values of K') leads to a minor increase in the computational overhead. Hence, it is feasible to moderately increase the number of neighbors retrieved by CAN-based search in order to achieve a higher accuracy and generate better predictions.

Since the precision of the heuristic CAN-based KNN search may seem low for small values of K', we conducted another two experiments, aimed at evaluating the quality of the neighborhood retrieved by the heuristic search. In the first, this was done by computing the average similarity between the nearest neighbors retrieved by the heuristic search and the active user. The computed average similarity was compared to the average similarity of neighborhood retrieved by the traditional search.

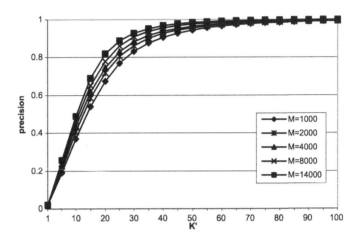

Fig. 6.6. Precision of CAN-based KNN.

In the experiment, we gradually increased the number of users inserted into the system from $M = 1000$ to $M = 14{,}000$. For each value of M, we compared the average similarity of heuristically retrieved neighbors with the average similarity of exhaustively retrieved neighbors for $K = K' = 10$. For each value of M, the above experiments were repeated 1000 times for different active users. The results of the experiment are shown in Fig. 6.7 (they are discussed after Fig. 6.8). The horizontal axis stands for the number of users inserted into the system, whereas the vertical reflects the average similarity value between the users in KNN set and the active user for both exhaustive and heuristic searches.

The second experiment was designed to evaluate the quality of the heuristically retrieved neighborhood by comparing the accuracy of the generated predictions. The final goal of the KNN search is to retrieve a set of the most similar users, whose ratings will be aggregated when generating the predictions. Thus, we generated the predictions using both exhaustively and heuristically retrieved sets of K-Nearest Neighbors and evaluated the accuracy of the predictions using well-known Mean Average Error (MAE)

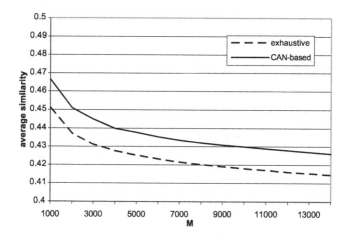

Fig. 6.7. Average similarity versus the number of users inserted.

metric:[6]

$$\text{MAE} = \frac{\sum_{i=1}^{N} |p_i - r_i|}{N} \quad (6.4)$$

where N denotes the number of predicted items, and p_i is the predicted, and r_i is the real rating on item i.

Also in this experiment the number of users inserted into the system was gradually increased from $M = 1000$ to $M = 14,000$. For each value of M, the experiment was repeated 1000 times for various, randomly chosen active users. For each active user chosen, the following operations were conducted: (1) a single randomly selected rating in the user's profile was hidden and served as a rating to be predicted, while the remaining *all-but-one* ratings served as the user's profile, (2) based on the all-but-one user's profile, the set of $K = K' = 10$ nearest neighbors was retrieved using both traditional exhaustive and heuristic retrievals, (3) predictions were generated using both heuristically and exhaustively retrieved neighborhoods, and (4) the MAE error of the generated predictions relatively to the original hidden rating was computed. The average values of the MAE computed for certain values of M are shown in Fig. 6.8. The horizontal axis stands for the number of users inserted into the system, whereas the vertical reflects the MAE values for both exhaustive and heuristic searches.

The results show that the average similarity (which is actually the dissimilarity) and the MAE of the predictions decrease with M. This is ex-

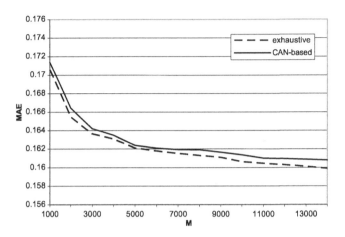

Fig. 6.8. Mean average error of the predictions versus the number of users inserted.

plained by the observation that the probability of discovering a similar user increases with the number of users inserted into the system. Thus, the average dissimilarity of the retrieved K-Nearest Neighbors decreases with M, while the accuracy of the generated predictions increases, and the MAE decreases as well.

Although both the similarity and the MAE of CAN-based heuristic search are higher (i.e. the retrieved neighbors are more dissimilar and the accuracy is actually lower), the curves are very close and the results are quite similar. Average deviation of the similarities is 2.93% and of the MAEs is only 0.38%. Note that the average deviation of the MAE is significantly lower than the average deviation of the similarities, as the generated predictions are barely affected by the changes in the retrieved neighborhoods. These experiments allow us to conclude that the proposed heuristic algorithm succeeds in both retrieving similar neighborhoods and generating accurate predictions.

6.5.3. *Inherent clustering*

One of the basic assumptions, that allows us to limit the heuristic search to users, located in close vicinity to the active user, is the *inherent clustering*. That means that the distance between two similar users is lower than the distance between two arbitrary users. Thus, the following experiment was designed to verify the property of inherent clustering in the underlying content-addressable space.

For this, we computed the average and the standard deviation of the similarity of the users located $R = 1$, 2, and 3 routing hops from the active user. The experiments were conducted for $M = 1000$, 2000, 4000, 8000 and 14,000 users inserted into the system. For each value of M, the experiments were repeated 1000 times for different random orders of inserting the users into the system and for different active users. Figure 6.9 shows the average similarity and the standard deviation as a function of R for the above values of M. The horizontal axis stands for M, the number of users inserted into the system, whereas the vertical reflects the average and the standard deviation of the similarity of the retrieved users, located within a given number of hops from the active user.

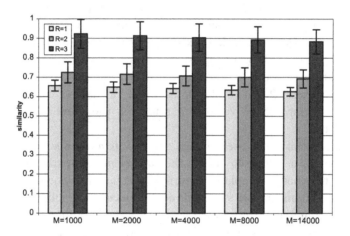

Fig. 6.9. Average similarity versus number of hops from the active user.

It can be seen that for any given value of M the similarity increases with R. This means that the similarity of users, located close to the active user is higher than the similarity of those located far. Thus, this experiment verifies our assumption on the clustering in content-addressable space. For any R, the average similarity and the standard deviation steadily decrease with M. This observation is explained by the fact that higher number of users leads to a better organization of zones, where zones managed by more similar users *block* the zones managed by dissimilar users. Thus, the average similarity (and the standard deviation) of users located within a given number of hops decreases with R.

Moreover, this experiment demonstrates the stability of the proposed CAN-based structure of users. This experiment was repeated 1000 times, for different random orders of inserting the users into the system. Low values of the standard deviation, and the steady decrease of it with the number of users in the system, show that the inherent clustering holds regardless of the different types of organization of the CAN zones, imposed by the different orders of inserting the users. Thus, we can conclude that the proposed heuristic KNN search will also succeed in retrieving accurate neighborhoods of users for different system usage scenarios.

6.5.4. *Completion heuristics*

The following experiments were designed to evaluate the proposed completion heuristics for filling-in the missing values in the incomplete ratings vectors. To run the experiment with the incomplete vectors, we used the full Jester dataset.[4] In previous experiments we used a partial dataset of complete vectors, built by 14,192 users that rated all 100 jokes. In addition, the full dataset also contains the ratings of 59,229 users that rated on average 45.26 jokes. The full Jester dataset (i.e. the dataset containing both complete and incomplete vectors) was used in the completion heuristics experiments.

We implemented the *user-average* and the *item-average* heuristics that were discussed in Sec. 6.4. As for the *conditional* heuristic, the decision regarding the chosen completion heuristic was based on the number of explicitly rated items in user's ratings vector. Since in the full Jester dataset the average number of items rated by a user was 45.26, in our implementation of the conditional heuristic the threshold for choosing an appropriate heuristic was set to 20 items. That means that if a user rated less than 20 items, his/her ratings vector is not considered as a reliable one, and vector completion exploits the *item-average* heuristic, which substitutes each missing rating with the average rating of the other users on the given item. However, if a user rated 20 items or more, the *user-average* heuristic is exploited, the missing ratings are substituted with the average rating of the given user on the other items.

To evaluate the accuracy of the proposed three completion heuristics, we conducted two types of experiments. In the first, we compared the average similarity value between the active user and the K-Nearest Neighbors retrieved by the heuristic search and by the traditional exhaustive search. The experiment was repeated three times, for the different completion heuristics exploited before inserting the completed vectors to the underlying content-addressable space.

In the experiment, we gradually increased the number of users inserted into the system from $M = 5000$ to $M = 50{,}000$. For each value of M, we compared the average similarity of the retrieved neighbors (using both exhaustive and heuristic retrieval techniques) for $K = K' = 10$. For each value of M, the above experiments were repeated 1000 times for different active users. The results of the experiment are shown in Fig. 6.10. The horizontal axis stands for M, the number of users inserted into the system, whereas the vertical reflects the average similarity value between the users in KNN set and the active user for both exhaustive and heuristic searches. Note that the heuristic retrieval was conducted three times, according to the slightly different datasets inserted into the content-addressable space, as imposed by the completion heuristics being exploited.

The curves show, that similarly to the accuracy results in previous subsections, the average similarity (i.e. dissimilarity) of the retrieved KNN users decreases with M, the number of users inserted into the system. Comparison of the proposed completion heuristics yields that the personalized *user-average* heuristic outperforms the nonpersonalized *item-average* heuristic. Average similarity deviation of the KNN set exploiting the *user-average* heuristic from the exhaustively retrieved KNN is 4.43%, while the similarity deviation of the *item-average* KNN set is 6.21%. Since the *conditional* heuristic is a smarter combination of the above heuristics, it slightly outperforms the *user-average* heuristic as well, and for it the average similarity deviation from the exhaustively retrieved KNN set is 4.11%.

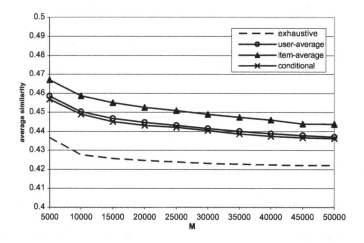

Fig. 6.10. Average similarity versus the number of users inserted.

Since the ultimate goal of the Collaborative Filtering is to generate predictions, the second experiment was designed to evaluate the quality of the completion heuristics by comparing the accuracy of the generated predictions. To do this, we generated the predictions using both exhaustively and heuristically retrieved sets of K-Nearest Neighbors and evaluated the accuracy of the predictions using the MAE metric. In the experiment the number of users inserted into the system was gradually increased from $M = 5000$ to $M = 50,000$. For each value of M, the experiment was repeated 1000 times for various, randomly chosen active users. The experimental setting was similar to one described in previous subsections: the user's profile was partitioned to the predicted rating and all-but-one profile, the sets of $K = K' = 10$ nearest neighbors were retrieved using both exhaustive and heuristic retrievals, the predictions were generated using both neighborhoods, and the MAE of the generated predictions relatively to the original rating was computed. The average values of the MAE are shown in Fig. 6.11. The horizontal axis stands for the number of users inserted into the system, while the vertical reflects the MAE values for both exhaustive and heuristic searches. Note that the heuristic retrieval was conducted three times, according to the completion heuristics being exploited.

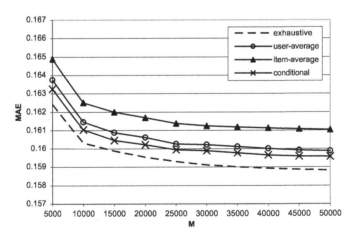

Fig. 6.11. Mean average error of the predictions versus the number of users inserted.

Similarly to the results in Sec. 6.5, this experiment shows that the MAE of the prediction decreases with M, the number of users inserted into the system. Comparison of the proposed completion heuristics yields that the accuracy of the predictions exploiting personalized *user-average* heuristic is better than that of the nonpersonalized *item-average* heuristic. However, for both heuristics the average increase of the MAE values is minor: for the *user-average* heuristic it is 0.69%, whereas for the *item-average* heuristic it is 1.37%. As can be seen from the chart, also in this experiment the *conditional* heuristic slightly outperforms both of them, as for the *conditional* heuristic the increase of the MAE is only 0.46%. Hence, out of the proposed three completion heuristics, the *conditional* heuristic retrieves the most similar KNN set and generates the most accurate prediction. This allows us to conclude that this heuristic should be used for converting the incomplete vectors to the complete ones, and naturally leads to future research, dealing with developing more accurate completion heuristics.

6.6. Conclusions and Future Research

One of the major drawbacks of the state-of-the-art CF implementations is their high computational complexity, which grows linearly both with the number of users and items in the system. In this work we proposed to heuristically decrease the required computational effort by implementing the CF over content-addressable CAN-like N-dimensional space.

6.6.1. *Conclusions*

Experiments conducted over the Jester dataset of jokes ratings show that in general the proposed heuristic algorithm outperforms the traditional exhaustive KNN search as the computational overheads are significantly decreased, while the accuracy remains roughly unchanged. Our algorithm decreases the number of required comparisons, while the ratio between the numbers of comparisons steadily decreases with the number of users. For example, for 14,000 users the number of comparisons was decreased by almost an order of magnitude (precisely, by 87%). Other experiments show that the number of comparisons roughly remains unchanged when K increases. This allows us to increase the number of nearest neighbors to be retrieved (and to potentially improve the accuracy of the generated predictions) with a very minor computational overhead.

In the accuracy experiments we qualitatively compared the neighborhoods retrieved and the predictions generated by the CAN-based heuristic and by the traditional exhaustive KNN searches. The retrieved neighborhoods were similar and the predictions were very close, which indicates good accuracy of the proposed algorithm. In summary, comparing the proposed heuristic KNN search with traditional exhaustive search shows that our algorithm achieves high accuracy (very similar to the accuracy of the traditional exhaustive KNN search), while significantly decreasing the required computational effort.

Another set of experiments were aimed at validating the inherent clustering property of content-addressable spaces. The results showed that this property holds in the CAN-like space, as the dissimilarity of users, located in a certain number of network hops from the active user increased with the number of network hops. The experiments also showed that the inherent clustering property holds regardless of the number of users inserted into the system and the order of their insertion.

The last set of experiments were aimed at comparing three heuristic for converting the incomplete vectors to complete ones by filling-in the missing ratings. Three simple heuristics were compared: two heuristics that substitute the missing ratings either with the average rating of the given user, or with the average rating on the given item, whereas the third heuristic integrates the first two. The experiments showed that the heuristic, which conditionally integrates two other heuristics, outperforms them both in terms of the retrieved neighborhoods' similarity and of the generated predictions' accuracy.

Comparing the MAE of the predictions generated by the complete and heuristically completed vectors yields that the accuracy of the predictions generated by the complete vectors is slightly better. This conclusion is reasonable, since the proposed completion heuristics insert some extent of noise into the original ratings. However, the increase in the MAE is minor, allowing us to conclude that the achieved computational optimization is preferential than the minor noises in the generated predictions caused by the artificial ratings inserted by the completion heuristics.

6.6.2. Future research

In this work, we inherently assumed that the system assigns equal relative weights to the ratings on each item. However, this assumption is not true in many real-life personalization applications. For example, this assumption might be false in a situation, where different criteria affect differently on the similarity values, e.g. when the similarity values between the items are known. Developing a weighted prediction algorithm will result in a more accurate Recommender System.

Also, we assumed that either the user's ratings on the items are available or they can be easily filled-in using one of the proposed simple completion heuristics. However, in some real-life scenarios, this completion is hard to achieve, since the matrix is very sparse (e.g. density of 2–3% in typical Collaborative Filtering datasets such as in Refs.[6] and[22] and the substitution of the missing values may require exploiting more intelligent techniques. In the future, we plan to study the use of various completion heuristics, exploiting statistical and Machine Learning techniques.

In addition to decreasing the computational effort, the proposed algorithm can naturally be extended to distribute it among multiple users. In traditional centralized implementations of the CF, the Similarity Computation and the Neighborhood Formation stages are performed in a single central location. However, as the underlying CAN platform is originally distributed Peer-to-Peer platform, it inherently allows distributed and fully decentralized storage of the ratings matrix. In the future, we plan to implement a distributed variant of the algorithm and to investigate the distribution issues.

The current work is limited to the Mean Squared Difference (MSD) similarity metric, since the injective mapping to a multidimensional CAN-like space inherently supports it. However, for other metrics, such as Cosine Similarity or Pearson Correlation, CAN space might be inappropriate and new types of topologies and respective mappings should be developed. We plan to study other metrics and to produce a general framework for efficient heuristic Collaborative Filtering.

Acknowledgments

The authors gratefully acknowledge the support of the Caesarea Edmond Benjamin de Rothschild Foundation Institute for Interdisciplinary Applications of Computer Science (CRI) and the Haifa Interdisciplinary Research Center for Advanced Computer Science (HIACS), both at the University of Haifa.

References

1. P. Resnick and H. R. Varian, Recommender systems, *Commun. ACM.* **40**(3), (1997).
2. B. Sarwar, G., Karypis, J. Konsta, and J. Riedl. Analysis of recommendation algorithms for e-commerce. In *Proc. EC Conf.*, (2000).
3. N. Good, J. B. Schafer, J. A. Konstan, A. Borchers, B. Sarwar, J. Herlocker,

and J. Riedl. Combining collaborative filtering with personal agents for better recommendations. In *Proc. AAAI Conf.*, (1999).
4. K. Goldberg, T. Roeder, D. Gupta, and C. Perkins, Eigentaste: a constant time collaborative filtering algorithm, *Inform. Retr. J.* **4**(2), 1331–151, (2001).
5. S. Aguzzoli, P. Avesani, and P. Massa. Collaborative case-based recommender system. In *Proc. ECCBR Conf.*, (1997).
6. J. L. Herlocker, J. A. Konstan, A. Borchers, and J. Riedl. An algorithmic framework for performing collaborative filtering. In *Proc. SIGIR Conf.*, (1999).
7. M. Morita and Y. Shinoda. Information filtering based on user behavior analysis and best match retrieval. In *Proc. SIGIR Conf.*, (1994).
8. F. Ricci, A. Venturini, D. Cavada, N. Mirzadeh, D. Blaas, and M. Nones. Product recommendation with interactive query management and twofold similarity. In *Proc. ICCBR Conf.*, (2003).
9. R. Burke, Hybrid recommender systems: survey and experiments, *User Modeling and User-Adapted Interaction.* **12**(4), 331–370, (2002).
10. U. Shardanand and P. Maes. Social information filtering: algorithms for automating word of mouth. In *Proc. CHI Conf.*, (1995).
11. J. Breese, D. Heckerman, and C. Kadie. Empirical analysis of predictive algorithms for collaborative filtering. In *Proc. UAI Conf.*, (1998).
12. S. H. S. Chee, J. Han, and K. Wang. Rectree: an efficient collaborative filtering method. In *Proc. DaWaK Conf.*, (2001).
13. S. Ratnasamy, P. Francis, M. Handley, R. Karp, and S. Shenker. A scalable content-addressable network. In *Proc. Conf.*, (2001).
14. D. M. Pennock, E. Horvitz, and C. L. Giles. Social choice theory and recommender systems: analysis of the axiomatic foundations of collaborative filtering. In *Proc. AAAI Conf.*, (2000).
15. A. Tveit. Peer-to-peer based recommendations for mobile commerce. In *Proc. WMC Workshop*, (2001).
16. B. M. Sarwar, J. A. Konstan, and J. Riedl. Distributed recommender systems: new opportunities for internet commerce. In *Internet Commerce and Software Agents: Cases, Technologies and Opportunities*, Idea Group Publishing, (2001).
17. B. N. Miller, J. A. Konstan, and J. Riedl, Pocketlens: toward a personal recommender system, *ACM Trans. Inform. Syst.* **22**(3), 437–476, (2004).
18. P. Han, B. Xie, F. Yang, and R. Shen, A scalable p2p recommender system based on distributed collaborative filtering, *Exp. Syst. Appl. J.* **27**(2), 203–210, (2004).
19. C. Plaxton, R. Rajaraman, and A. Richa. Accessing nearby copies of replicated objects in a distributed environment. In *Proc. ACM SPAA Conf.*, (1997).
20. S. Bogaerts and D. Leake. Facilitating cbr for incompletely-described cases: distance metrics for partial problem descriptions. In *Proc. ECCBR Conf.*, (2004).

21. G. Salton and M. McGill, *Introduction to Modern Information Retrieval.* (McGraw-Hill Publishers, 1983).
22. P. McJones. Eachmovie collaborative filtering data set. In *available online at http://research.compaq.com/SRC/eachmovie/,* (1997).

BIOGRAPHIES

Shlomo Berkovsky is a Ph.D. student in the Computer Science and Management Information Systems Departments at the University of Haifa. He received his M.Sc. degree from the Computer Science Department of the University of Haifa in 2004 in the area of semantic data management in peer-to-peer networks. He has published over 20 papers in scientific journals and conferences.

His research focuses on user models mediation for a better personalization in recommender systems.

Yaniv Eytani is a Ph.D. candidate in the Computer Science Department at the University of Illinois at Urbana-Champaign and is involved in research under the direction of Prof. Grigore Rosu. He is a member of the Formal Systems Laboratory. He received an M.Sc. degree from the Computer Science Department of the University of Haifa in 2005 in the field finding concurrent bugs in Java. He has published about 20 papers in scientific journals and conferences.

Larry M. Manevitz received his B.Sc. degree from Brooklyn College, and his M.Phil. and Ph.D. degrees from Yale University, Mathematics Department in Mathematical Logic. He has held positions at Hebrew University, Bar-Ilan University, Oxford University, NASA Ames, University of Texas at Austin, Baruch College, CUNY, University of Maryland and University of Wisconsin. Currently he is in the Department of Computer Science, Univ. Haifa, and is the Director of the HIACS Research Laboratory and the Neurocomputation Laboratory there. He has about 60 scientific publications.

He currently specializes in artificial neural networks, machine learning, and brain modeling and has done substantial work in both theoretical and applied mathematical logic (especially non-standard analysis) and in the theory of combining uncertain information.

Chapter 7

Identifying and Analyzing User Model Information from Collaborative Filtering Datasets

Josephine Griffith* and Colm O'Riordan* and Humphrey Sorensen[†]

*Department of Information Technology
National University of Ireland, Galway, Ireland
josephine.griffith@nuigalway.ie
colmor@it.nuigalway.ie
† Department of Computer Science
University College Cork, Ireland
sorensen@cs.ucc.ie

This paper considers the information that can be captured about users from a collaborative filtering dataset. The aims of the paper are to create a user model and to use this model to explain the performance of a collaborative filtering approach. A number of user features are defined and the performance of a collaborative filtering system in producing recommendations for users with different feature values is tested. Graph-based representations of the collaborative filtering space are presented and these are used to define some of the user features as well as being used in a recommendation task.

7.1. Introduction

Modern information spaces are becoming increasingly more complex with information and users linked in numerous ways, both explicitly and implicitly, and where users are no longer anonymous, but generally have some identification and a context in which they navigate, search and browse. This offers new challenges to recommender system designers, in capturing and combining this information to provide a more personalized and effective retrieval experience for a user.

The original foundations of collaborative filtering came from the idea of "automating the word of mouth process" that commonly occurs within social networks,[1] i.e. people will seek recommendations on books, CDs, restaurants, etc. from people with whom they share similar preferences in these areas.

Although collaborative filtering is most frequently seen as a way to provide recommendations to a set of users, collaborative filtering datasets also allow for the analysis of social groups and of individual users within a group, thus providing a means for creating a new user model, group model or for augmenting an existing user or group model.

User modelling has had a long history in many computer science domains and traditionally user models were created based on evidence from explicit user actions. There has been a gradual change in this approach and the focus is often on building a model for a user using implicit information gleaned from the user's interactions with a system, the user's interactions with data and information, and the user's interactions with other users.

A social network can be defined as a graph representing relationships and interactions among individuals.[2] Nodes in the graph represent individuals and the links between the nodes represent some relationship or relationships between individuals. Many modern social networks are found on the Internet in the form of virtual communities and the study and analysis of social networks occur in many different fields. A number of systems based on social networks and small world networks have been proposed for referral and recommendation.[3-7] Other work linking social networks and collaborative filtering has viewed the collaborative filtering dataset as a social network with the aim of analyzing properties of users and items to improve retrieval performance.[8-11] Aims other than solely improving retrieval performance have also been explored.[9]

This paper considers the ways that recommender systems bring users together and considers how the information from these recommender systems can be extracted to form user models. The motivation for this work is that although, in collaborative filtering approaches, users are often clustered into groups based on finding "similar users", there is no modelling of the features of a particular user or group. Also, with the exception of simple cases (e.g. when a user has given very few ratings), it is not clear what effect these features have on recommendation accuracy.

The goals of the work presented in this paper are to specify some of the information that can be captured about users given a collaborative filtering dataset and to provide a model that will represent these features of users. In this work, eight features that can be extracted from the collaborative filtering dataset are firstly identified and defined. Some of these features are particular to the recommendation task while some features use measures from social network theory and information retrieval. The eight features are then analyzed with respect to their effect on recommendation accuracy using a collaborative filtering approach. This is done, for each feature, by taking sample test users that have a particular value for the feature and by testing the accuracy of a collaborative filtering recommender system in providing predictions for the test users.

The user model defined will be used in future work to ascertain if improvement in recommendation accuracy can be achieved (by allowing the development of more personalized recommender algorithms) and also the model will be used to maintain histories of users in a collaborative filtering information space.

Section 7.2 presents related work in collaborative filtering, graph-based approaches to recommendation and social networks. Section 7.3 outlines the methodology, presenting the collaborative filtering approach and the graph models used as well as specifying the user features which are extracted from the collaborative filtering dataset. Section 7.4 discusses the experiments performed and the experimental set-up. Section 7.5 presents results and Sec. 7.6 presents conclusions, discussing the potential usefulness of the features and approach and outlining future work.

7.2. Related Work

Given a set of users, a set of items, and a set of ratings, collaborative filtering systems attempt to recommend items to users based on user ratings. Collaborative filtering systems generally make use of one type of information, that is, prior ratings that users have given to items. However, some recent work has investigated the incorporation of other information, for example, content,[12] time,[13] and trust[14] information. To date, application domains have predominantly been concerned with recommending items for sale (e.g. movies, books, CDs) and with small amounts of text such as Usenet articles and email messages. The datasets within these domains will have their own characteristics, but they can be predominantly distinguished by the fact that they are both large and sparse, i.e. in a typical domain, there are many users and many items but any user would only have ratings for a small percentage of all items in the dataset.

The problem space can be viewed as a matrix consisting of the ratings given by each user for the items in a collection, i.e. the matrix consists of a set of ratings $r_{a,i}$, corresponding to the rating given by a user a to an item i. Using this matrix, the aim of collaborative filtering is to predict the ratings of a particular user, a, for one or more items not previously rated by that user. The problem space can equivalently be viewed as a graph where nodes represent users and items, and nodes and items can be linked by weighted edges in various ways. Graph-based representations have been used for both recommendation and social network analysis of collaborative filtering datasets.[11,15]

7.2.1. Weighting schemes in collaborative filtering

There has been much work undertaken in investigating weighting schemes for collaborative filtering where these weighting schemes typically try to model some underlying bias or feature of the dataset in order to improve prediction accuracy. For example, in Ref. 16 and Ref. 17 an inverse user frequency weighting was applied to all ratings where items that were rated frequently by many users were penalized by giving the items a lower weight. In Ref. 18 and Ref. 17 a variance weighting was used which increased the influence of items with high variance and decreased the influence of items with low variance. The idea of *tf-idf* weighting scheme from information retrieval was used in Ref. 19 (using a row normalisation) and in Ref. 20 (using a probabilistic framework). Work in Ref. 21, Ref. 22 and Ref. 11 involve learning the optional weights to assign to items. In Ref. 14 more weight is given to user neighbours who have provided good recommendations in the past (this weight is calculated using measures of "trust" for users) and in Ref. 23 more weight is given to items which are recommended more frequently (where the weights are calculated using an "attraction index" for items). In general, although some of the weighting schemes for items have shown improved prediction accuracy (in particular those involving learning), it has proven difficult to leverage the feature information to consistently improve results.

7.2.2. Graph-based approaches for recommendation

Several researchers have adopted graph representations to develop recommendation algorithms. A variety of graphs have been used, including, among others, directed, two-layer, etc. and a number of graph algorithm approaches have been adopted (e.g. horting,[24] spreading activation[15]).

Aggarwal *et al.* present *horting*, a graph-based technique where nodes represent users and directed edges between nodes correspond to the notion of predictability.[24] Predictions are produced by traversing the graph to nearby nodes and combining the ratings of the nearby users.

Huang et al. present a two-layer graph model where one layer of nodes corresponds to users and one layer of nodes corresponds to items.[15] Three types of links between nodes are represented: item–item links representing item similarity based on item information, user–user links representing user similarity based on user information, and inter-layer user–item links between items and users that represent a user's rating (implicit or explicit) for an item. Transitive relationships between users, using a subset of this graph representation, are explored in Ref. 25. A bipartite graph is used with one set of nodes representing items and the second set of nodes representing users. Binary weighted edges connect the nodes between the two sets where an edge has a weight of 1 if a user purchased, or gave positive feedback to, an item and a weight of 0 otherwise. The goal is to compare how well different collaborative filtering approaches deal with the sparsity problem and the cold start problem for new users.

A number of approaches have been proposed to effect retrieval and filtering using graph representations. One such approach is spreading activation which originated from the field of psychology and was first used in computer science in the area of artificial intelligence to process semantic networks. Spreading activation approaches have been used in many information retrieval applications[26] and more recently in the domain of collaborative filtering.[25] Spreading activation approaches have also been used to integrate sources of evidence and information.[27,28]

7.2.3. *Collaborative filtering as a social network*

As well as being used for recommendation, a collaborative filtering dataset has been viewed as a social network where nodes in the network represent users and the links between users can be calculated based on the items users have accessed and/or the actual ratings that users have given to these items.[2,10,11] Rashid et al. state that "In contrast to other social networks, recommender systems capture interactions that are *formal, quantitative, and observed*."[11]

A social network can be defined as a network (or graph) of social entities (e.g. people, markets, organizations, countries), where the links (or edges) between the entities represent social relationships and interactions (e.g. friendships, work collaborations, social collaborations, etc.). Recently, online relationships between people have also been used to create social networks.

A number of recommender systems based on social networks and small world networks have been developed. Such social networks have been built using histories of email communication,[6] co-occurrence of names on WWW pages,[3] co-use of documents by users,[29] and matching user models and profiles.[7]

Palau et al. represent the agents in a multiagent restaurant recommender system using a social network where the connections between agents are based on the level of trust the agents have in the recommendations of other agents. Social network theory measures of size, density, network centrality, and clique and faction substructures are used to help give an explanation of the performance of the system.[10]

Lemire considers the social network feature of *influence* and found that recommendation results were better if the system was not "too democratic", i.e. it was found that it was better not to penalize users with a high number of ratings.[8] In addition, Lemire discusses the *stability* of a collaborative filtering system, defining stability as a property which exists if a single user in a large set does not make a difference to the results for some active user.

Mirza et al. also induce a social network from a collaborative filtering dataset where connections between users are based on the co-ratings of the same items.[9] They define a *hammock jump* as a connection between two users in the network that will exist if the users have co-rated at least w items (where w is defined as the hammock width). Herlocker et al. refer to this measure as a *significance weighting* whereby they devalue the correlation value between two users if this correlation value has been calculated based on only a small number of co-rated items.[30]

In Ref. 31, a graph-based representation is used to analyze various features of a dataset in order that the suitability of a collaborative filtering algorithm to a particular dataset can be ascertained (in particular, to give an indication of whether a naive (Top-N), user–user, item–item or spreading activation collaborative filtering algorithm would work well with the dataset).

7.3. Methodology

In this paper the focus is to extract implicit user information available from the collaborative filtering dataset and to form a user model for each user. This implicit information is based on simple features which can be extracted from any recommendation dataset (e.g. number of items rated, liked, disliked, etc.) as well as extracting features which

Identifying and Analyzing User Model Information from Collaborative Filtering 171

are based on measures from social network theory and information retrieval.

The user model consists of a octet containing, for each user, the values for eight identified features. Each of the eight features is individually analyzed by considering a set of users with different values for the feature. For example, one set of users are those who have rated close to the maximum number of items rated; another set of users are those who have rated close to the average number of items rated. Test users from each of these sets are chosen as the active users of a collaborative filtering recommender system, i.e. these are the users for which a recommendation is sought. The accuracy of the recommender system in providing recommendations for each set of users is found and these results are compared for each feature (e.g. the set of users who rated close to the maximum number of items may receive better predictions than the set of users who rated close to the minimum number of items). Therefore, for an individual user, some explanation as to why the system performs poorly or well for the user can be given by looking at the feature values for that user.

In addition we investigate the performance of a graph-based approach to recommendation with a view to incorporating the user features into a graph representation in future work. A graph-based representation is also used to define some of the user features.

The collaborative filtering system that is used to provide recommendations to different sets of users is described in Sec 7.3.1. Two graph-based representations of the collaborative filtering problem are presented in Sec 7.3.2. Each of the eight features of the user model are defined in Sec 7.3.3.

7.3.1. *Collaborative filtering approach*

The collaborative filtering problem space is often viewed as a matrix consisting of the ratings given by each user for some of the items in a collection. Using this matrix, the aim of collaborative filtering is to predict the ratings of a particular user, a, for one or more items not previously rated by that user. Memory-based techniques are the most commonly used approach in collaborative filtering although numerous other approaches have been developed and used.[16] Generally, traditional memory-based collaborative filtering approaches contain three main stages (for some active user a):

(1) Find users who are similar to user a (the neighbours of a).
(2) Select the "nearest" neighbours of a, i.e. select the most similar set of

users to user a.

(3) Recommend items that the nearest neighbours of a have rated highly and that have not been rated by a.

Standard statistical measures are often used to calculate the similarity between users in step 1 (e.g. Spearman correlation, Pearson correlation, etc.).[32] In this work, similar users are found using the Pearson correlation coefficient formula (7.1):

$$\text{corr}_{a,u} = \frac{\sum_{i=1}^{m} (r_{a,i} - \bar{r}_a) \times (r_{u,i} - \bar{r}_u)}{\sqrt{\sum_{i=1}^{m} (r_{a,i} - \bar{r}_a)^2} \times \sqrt{\sum_{i=1}^{m} (r_{u,i} - \bar{r}_u)^2}} \quad (7.1)$$

where $\text{corr}_{a,u}$ is the correlation value between users a and u (a value in the range $[-1, 1]$) for m items rated by users a and u, $r_{a,i}$ is the rating given by user a to item i, $r_{u,i}$ is the rating given by user u to item i, $\bar{r}_a$ and $\bar{r}_u$ are the average ratings given by users a and u, respectively.

An adjustment (using a *significance weighting*) is used in the Pearson correlation calculation based on the number of items that users have rated in common (co-rated items).[30] The motivation is that two users may receive a high correlation value but this might only be based on a small number of co-rated items. The adjustment ensures that users must have similar preferences over more than a few items to be considered highly correlated. The adjustment used in this work involves multiplying $\text{corr}_{a,u}$ by the significance weighting if the number of co-rated items is less than twice the average number of co-rated items. The significance weighting between two users a and u is defined as (7.2):

$$\frac{\text{cr}_{a,u}}{2 \times average} \quad (7.2)$$

where $\text{cr}_{a,u}$ is the number of items users a and u have co-rated and *average* is the average number of items that have been co-rated by all users in the dataset.

The "nearest" neighbors of a user are selected using a low neighbour selection threshold, with any correlation value greater than 0.01 being considered. Although Breese[16] found that users with high correlation values (> 0.5) were more valuable in providing recommendations, work by Herlocker[33] using the Movie Lens dataset found that, for this dataset, such a high threshold sacrificed coverage and in addition, higher thresholds never improved the accuracy of predictions. They found that experiments using all correlation values greater than 0 always outperformed experiments with higher thresholds.[33]

Rating predictions for items for a user a (step 3) are found using the formula (7.3):

$$\text{pred}_{a,i} = \frac{\sum_{u=1}^{n}(r_{u,i} - \bar{r_u}) \times \text{corr}_{a,u}}{\sum_{u=1}^{n} \text{corr}_{a,u}} \qquad (7.3)$$

where $\text{pred}_{a,i}$ is the prediction for item i for user a, n is the number of neighbors of user a, $r_{u,i}$ is the rating given by user u to item i, $\bar{r_u}$ is the average rating given by user u, and $\text{corr}_{a,u}$ is the correlation value between users a and u.

7.3.2. Graph-based representations of the collaborative filtering space

As discussed in the previous section the collaborative filtering problem space is often viewed as a matrix. The problem space can equivalently be viewed as a graph consisting of a set of user nodes and a set of item nodes. Two different graph representations are considered in this work.

In the first representation (see Fig. 7.1), user and item nodes are connected via weighted edges where the weights on the edges represent the ratings given to items by users. Apart from some scaling of the rating values this graph is a direct mapping of the matrix representation of the data to a graph representation of the data.

The second representation (see Fig. 7.2) is a social network representation which only considers user nodes. These user nodes are connected via weighted edges if the users are deemed sufficiently similar to each other. This similarity is calculated using the Pearson correlation formula where positive correlation values indicate similarity. A threshold value of 0.25 is used so that an edge only exists between users if their correlation value is greater than 0.25. For the collaborative filtering case, commonly used correlation measures are not commutative so therefore in the representation used, two edges can exist between two users.

Note that the two representations can be combined into a single graph representation. Currently, in the given representations, the information on user features is not represented explicitly. To represent this information explicitly, additional edges can be added to the graph to represent further relationships between users and items, relationships between items and relationships between users. For example, a relationship can exist between commonly rated items; between highly rated items, etc.

To provide recommendations, the graph representation in Fig. 7.1 is augmented such that three weighted edges connect nodes: one undirected

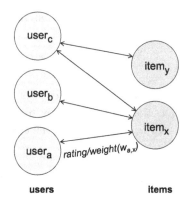

Fig. 7.1. Graph representation of users, items and ratings.

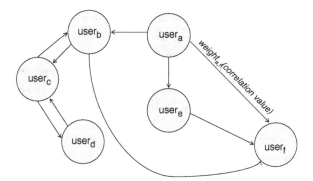

Fig. 7.2. Graph representation of users and their similarity.

edge representing the rating (or weight, w_i) and the second and third directed edges representing node outputs (output_i). Associated with each user node and item node is an activity and a threshold (see Fig. 7.3 which shows this augmentation for a portion of the graph from Fig. 7.1).

The activity of a user or item node a, for N nodes connected to the node a with nonzero weight, is calculated by (7.4):

$$\text{activity}_a = \sum_{i=1}^{N} \text{output}_i w_i \qquad (7.4)$$

where output_i is the output of the node i that is connected to node a and w_i is the weight on the edge connecting node i to a. The output, output_a,

Identifying and Analyzing User Model Information from Collaborative Filtering 175

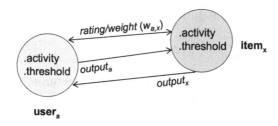

Fig. 7.3. Extended graph representation of a single user and item node.

of a user or item node is calculated using a threshold function (7.5):

$$\text{output}_a = \begin{cases} \text{activity}_a & \text{if activity}_a > \tau, \\ 0 & \text{otherwise}, \end{cases} \quad (7.5)$$

where the threshold function uses the node's activity and a threshold value, τ. Each node may have its own threshold value.

The terminology of a hop is used in this paper to define the activation spreading from one set of nodes to a second set of nodes. A hop involves the calculation of all node outputs in either the user set or item set, updating the associated activities and outputs of the nodes. The steps involved in the spreading activation approach are as follows:

1. Hop 1: Calculate the activities of all item nodes connected, with nonzero weight, to the current active user node. For each activated item node, calculate the output of the node using the threshold function.
2. Hop 2: Calculate the activities of all user nodes connected, with nonzero weight, to item nodes where the item nodes have nonzero output. For each activated user node, calculate the output of the node using the threshold function.
3. Hop 3: Calculate the activities of all item nodes connected, with nonzero weight, to user nodes where the user nodes have nonzero output. For each activated item node, calculate the output of the node using the threshold function.
4. Following three hops, items with the top-N highest positive activities are recommended to the active user.
5. Steps 2 and 3 can be repeated any number of times before recommendations are given (step 4).

Two hops result in activating a set of user nodes constituting a user neighborhood of the original active user node. The third hop, from user nodes to item nodes, provides item recommendations for the active user.

7.3.3. User features

A user model is defined which consists of eight features. For some user a the features are defined as follows:

(1) *rated* is the number of items rated by the user a.
(2) *liked* is the percentage of items rated by the user a that the user a liked, and is calculated by (7.6):

$$\frac{\text{num}_{\text{liked}}}{\text{rated}} \qquad (7.6)$$

where $\text{num}_{\text{liked}}$ is a count of the number of items liked by the user a and *rated* is the number of items rated by the user a. In this work, an item is considered to be liked by a user if it receives a value greater than the middle value of the rating range, as also used in Ref. 34 (e.g. if the rating range is [1, 5] a liked item is an item that receives a value of 4 or 5).

(3) *disliked* is the percentage of items rated by the user a that the user a disliked and is calculated by (7.7):

$$\frac{\text{num}_{\text{disliked}}}{\text{rated}} \qquad (7.7)$$

where $\text{num}_{\text{disliked}}$ is a count of the number of items disliked by the user a and *rated* is the number of items rated by the user a. An item is considered to be disliked by a user if it receives a value less than the middle value of the rating range.

(4) *avg-rating* is the average rating value given to items by the user a.
(5) *std-dev* is the standard deviation of the ratings of user a.
(6) *influence* is a measure of how influential a user is in comparison to other users. As also considered in Refs. 11 and 9, *influence* is defined in this work by using measures from social network theory. In particular, degree centrality is used where the dataset is viewed as a graph (or social network) where nodes represent users and the values of weights on edges between users are based on the strength of similarity of users to each other (as shown in Fig. 7.2). Degree centrality is then measured by counting the number of edges a node has to other nodes. Essentially this is a count of the number of neighbours (above a correlation threshold of 0.25) a user has.

(7) *clustering-coeff* is also a measure taken from social network theory and measures how similar users in a group are to each other using the clustering coefficient measure. This measures how connected the neighbours of the user a are to each other using the graph representation in Fig. 7.2. For example, if none of user a's neighbours are connected to each other, the clustering coefficient is 0 whereas if this subgraph has a clustering coefficient of 1 then all of user a's neighbours are connected to each other. The clustering coefficient is calculated by (7.8):

$$\frac{actual}{possible} \quad (7.8)$$

where *actual* is the number of actual links between neighbour nodes and *possible* is the number of possible links which can exist between neighbour nodes. In the representation described the total number of possible links that can exist between n nodes is $(n^2 - n)$.

In addition, in the collaborative filtering case it is possible that small sub-groups (small values of n) will have high clustering coefficients and therefore comparisons using clustering coefficient values may not always be meaningful. To overcome this the formula is extended to also include the active user in the calculation.[35] Thus the formula for the clustering coefficient for a user a with degree, $deg(a)$, and n neighbour nodes with degree greater than 1 becomes (7.9):

$$\frac{actual + deg(a)}{(n+1)^2 - n + 1} \quad (7.9)$$

Considering the graph shown in Fig. 7.2 with the active user being $user_a$, who has degree 3 and three neighbours ($n = 3$) who are connected to each other as follows: user e is connected to user f and user b is connected to user f. Therefore the number of actual links is 2 and the clustering coefficient for this group is 0.42.

(8) *importance* is a measure taken from Information Retrieval. Some collaborative filtering weighting schemes incorporate the idea from Information Retrieval of a *term frequency, inverse document frequency (tf-idf)* weighting.[19,20] The idea is to find terms with high discriminating power, i.e. terms which "describe" the document well and also distinguish it from other documents in the collection. Mapping the idea of *tf-idf* to collaborative filtering, a "term" can be viewed as a user with associated ratings for M distinct items. The more ratings a user has the more important the user is, unless the items that the user has rated have been rated frequently in the dataset. Note that the value a user

gives an item is not a frequency or weight - it is an indication that the item has been rated and thus the actual rating value is not used in the following formula 7.10. The formula used to calculate the importance, w_i, of a user i is:

$$w_i = \frac{1}{M} \times \sum_{j=1}^{M} \left(1 + \log \frac{n}{n_j}\right) \qquad (7.10)$$

where n is the total number of users in the dataset; M is the number of ratings by user i and n_j is the number of users who rated item j.

7.4. Experiments

This section presents details of the experiments performed using the collaborative filtering approach and the graph-based approach outlined previously. The first set of experiments involve analyzing each of the eight features identified in the previous section using a collaborative filtering approach. The final experiment involves testing the performance of the graph-based representation illustrated in Fig. 7.3 using a spreading activation approach to collaborative filtering.

7.4.1. *User model features*

The main experiments involve checking the relative performance of a collaborative filtering approach using different sets of users for each of the eight features. A set of users consists of the users who have the same value, or nearly the same value, for an identified feature. The aim is to ascertain which sets of users will be more likely to have better or worse predictions (measured using the mean absolute error (MAE) metric).

A standard subset of the Movie Lens dataset is considered that contains 943 users and 1682 movies. A proportion of the dataset is removed for testing, as described below, and the metric of mean absolute error is used to compare the performance of the collaborative filtering approach using different sets of users, for each feature, with different feature values.

For each feature, the range of values for that feature (e.g. [0.12, 1] for the *liked* feature) is broken into regular intervals (typically 8 intervals) and users belong to a particular interval based on their value for that feature. All users in a particular interval then form a set. Intervals are chosen such that the set size (the number of users in each interval) is close to 100.

Identifying and Analyzing User Model Information from Collaborative Filtering 179

For testing, 30 users are chosen randomly from each set as the test users and 10% of their ratings for items are removed to yield the items to test (i.e. the system should return predictions for these items). MAE results are averaged over 10 runs for each set of users, for each feature. In addition, for each feature a control set of 30 users is chosen randomly from the entire dataset as test users (i.e. the users are chosen without considering the feature value of these users).

7.4.2. *Spreading activation*

The purpose of this experiment is to test whether the graph representation and spreading activation approach are sufficiently accurate to be used in future work involving the incorporation of the user features into the graph representation. The experiment involves the comparison of a spreading activation approach and a traditional memory-based collaborative filtering approach. The reason for choosing a traditional memory-based collaborative filtering approach is that it has been shown to perform well in comparison to many other collaborative filtering techniques.[33,36] Also, as can be seen from the descriptions of the approaches in Secs. 7.3.1 and 7.3.2, it is quite similar to the spreading activation approach outlined. The important difference between the representations and approaches is in terms of the flexibility of the graph-based representation and spreading activation approach in allowing the incorporation of additional information.

Again, the Movie Lens dataset is used. Weights on the network edges indicate the strength of like/dislike for an item where "dislike" can be viewed as an inhibitory or negative rating and "like" can be viewed as an excitatory or positive rating. Given that the original rating values in the Movie Lens dataset are all positive numbers, the approach adopted maps the ratings to positive and negative values to indicate positive and negative influences. The mapping chosen is to subtract 2.5 from all nonzero values which will give ratings around 0, giving:

$$\{0, 1, 2, 3, 4, 5\} \rightarrow \{0, -1.5, -0.5, 0.5, 1.5, 2.5\}.$$

The value 0 is not changed in this mapping as it has a special meaning, being used to indicate that no rating has been given to an item.

A proportion of the dataset is removed for testing and the metric of precision is used to compare the performance of the two approaches at different recall points. Precision is used because the spreading activation approach returns a ranking of recommended items, not prediction values that can be compared with actual values.

The collaborative filtering approach and settings used are those already described. In the spreading activation approach to collaborative filtering, three stages corresponding to the three stages in the traditional memory-based collaborative filtering approach are used. Neighbours of some active user are found after two hops of the approach, at which stage user nodes that have nonzero activity are the neighbours of the active user. When activation is spread again, from user nodes to item nodes, items not rated by the active user will be highlighted. These items are recommended to the user if the activity is sufficiently high. The threshold value used in these experiments is 0 for all nodes, i.e. all positive activities will result in a node outputing a value.

7.5. Results

7.5.1. User model features

Results are presented for each of the eight features using the experimental methodology outlined in the previous section.

Figure 7.4 shows the MAE results when the *rated* feature was analyzed for eight sets of users. The *rated* value ranges from 0 to 737. The users in the first set (0–24 interval) have rated 0–24 items; the users in the second set (25–30 interval) have rated 25–30 items, etc. A random group of 30 users (with varying *rated* values) was also chosen (and are not included on the graph). This random group had an average MAE value of 0.7624. As expected, the worst MAE value for any set was for the users in the set who have rated between 0 and 24 items, i.e. these users have provided the very minimum number of ratings. Although we would expect that the accuracy should steadily increase as the number of ratings users have given increases, this was not necessarily the case. However, users who have rated closer to the maximum number of items have the best MAE values.

Figures 7.5 and 7.6 show the MAE results when the *liked* and *disliked* features were analyzed for nine sets of users. For the *liked* feature, values range from 0.12 to 1 where a value of 1 indicates that a user liked all the items that they rated and a value of 0.12 indicates that a user liked very few of the items that they rated. A random group of 30 users (with varying *liked* values) was also chosen and had an average MAE value of 0.7495. For the *disliked* feature values range from 0 to 0.87 where a value of 0 indicates that a user liked all the items that they rated. A random group of 30 users

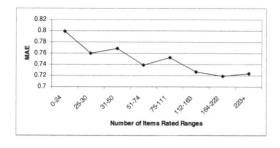

Fig. 7.4. *rated* MAE analysis.

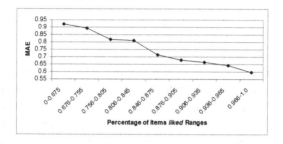

Fig. 7.5. *liked* MAE analysis.

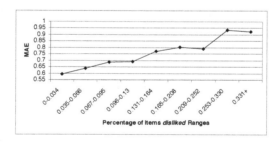

Fig. 7.6. *disliked* MAE analysis.

(with varying *disliked* values) were also chosen and had an average MAE value of 0.7507. As can be seen from both graphs, the results improve when the percentage of positively rated items (i.e. those liked by a user) increases.

Figure 7.7 shows the MAE results when the *avg-rating* feature was analyzed for eight sets of users. The *avg-rating* value ranges from 1.0 to 4.869. The MAE for 30 randomly chosen users was 0.7321. The users with lowest averages (from the minimum to 3.03) have the worst MAE and the users with the highest averages (> 4.10) have the best MAE.

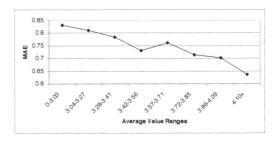

Fig. 7.7. *avg-rating* MAE analysis.

Fig. 7.8 shows the MAE results when the standard deviation feature (*std-dev*) was analyzed for eight sets of users. The *std-dev* value ranges from 0.3499 to 1.718. The users with low standard deviation (< 0.779) exhibited the best MAE value (0.5595 in comparison to the MAE of the randomly selected group which was 0.7779) while the users with the highest standard deviation had the worst MAE. This suggests that better recommendations can be found for users with lower variance in their ratings.

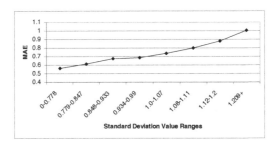

Fig. 7.8. *std-dev* MAE analysis.

Fig. 7.9 shows the MAE results when the *influence* feature was analyzed for eight sets of users. The *influence* value ranges from 0 to 392 where an *influence* value of 0 means that a user has no neighbours. As expected, the users with fewest neighbours (0 or 1) have the worst MAE values and as the neighbourhood size grows there is a general trend towards lower MAE values. The MAE of the random group was 0.7508.

The clustering coefficient feature (*clustering-coeff*) was analyzed for eight sets of users with values ranging from 0 to 0.864 where a value of

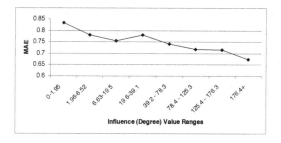

Fig. 7.9. *influence* MAE analysis.

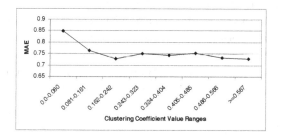

Fig. 7.10. *clustering-coeff* MAE analysis.

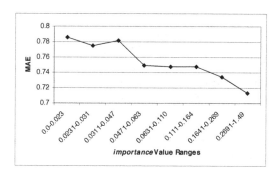

Fig. 7.11. *importance* MAE analysis.

0 means that none of the active user's neighbours are linked to each other (with a correlation value above 0.25). The MAE of the random group was 0.7479. The graph (Fig. 7.10) shows that as the *clustering-coeff* value increases towards 1 (i.e. the active user's neighbours are more similar to each other) the prediction accuracy very slightly improves. The poorest results

are seen for users who have very low clustering coefficient values.

The importance feature (*tf-idf*) was also analyzed for eight sets of users with values ranging from 0.015 to 1.485 (see Fig. 7.11). Results are poorer when a user has a low importance (*tf-idf*) value and results are better when a user has a high importance value. The MAE of the random group was 0.7255.

7.5.2. *Spreading activation*

Figure 7.12 illustrates the precision recall graph for the spreading activation approach and the traditional memory-based approach to collaborative filtering. Results were averaged over 100 runs. It can be seen that the spreading activation approach outperforms the traditional memory-based approach at all recall points other than the first. These results were shown to be statistically significant using a 2-tailed paired t-test at p-values < 0.05.

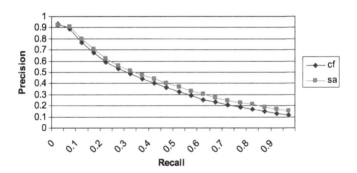

Fig. 7.12. Comparing spreading activation (sa) and traditional memory-based approaches (cf) to collaborative filtering.

This suggests that the graph-based representation and spreading activation approach give as good (and slightly better) performance as a traditional memory-based approach which has been shown to perform well. The advantage of the graph-based representation and spreading activation approach over other representations and approaches is their flexibility in allowing the incorporation of additional information. Given these results, future work can proceed in using the graph-based representation and spreading activation approach to incorporate information on user models.

7.6. Conclusions and Future Work

In this paper we have reviewed work in collaborative filtering, social networks and graph-based recommendation, highlighting the similarities between the work. We have defined a user model containing eight features of users that can be identified from a collaborative filtering dataset. We have shown how the prediction accuracy of a traditional memory-based collaborative filtering approach varies depending on the value of these features for certain users. This provides a first step at more personalized recommendations for users by providing some explanation for the relative good or poor performance of the collaborative filtering system (based on the values that users have for the identified features).

We have also shown some initial experimental evaluation of the usefulness of a graph-based representation of the collaborative filtering space using a spreading activation approach for recommendation.

We believe that more personalized and accurate recommendations can be obtained by incorporating the features identified in this paper into the graph models presented. Future work will explore these and other user features in more detail and will also consider the combination of these feature values. In addition, future work involves demonstrating that a graph-based representation of the collaborative filtering space allows the incorporation of these features and also other information on users, items and groups. This will strengthen the case for the application of graph-based recommendation algorithms.

References

1. U. Shardanand and P. Maes. Social information filtering: Algorithms for automating word of mouth. In *Proceedings of the Annual ACM SIGCHI on Human Factors in Computing Systems (CHI '95)*, pp. 210–217, (1995).
2. J. Barnes, *Social Networks*. (MA: Addison-Wesley, 1972).
3. H. Kautz, B. Selman, and M. Shah, Referral web: combining social networks and collaborative filtering, *Communications of the ACM.* **40**, 63–65 (March, 1997).
4. B. Krulwich and C. Burkey. The contactfinder: Answering bulletin board questions with referrals. In *Proceedings of the Thirteenth National Conference on Artificial Intelligence*, (1996).
5. D. McDonald and M. Ackerman. Expertise recommender: a flexible recommendation system and architecture. In *Proceedings of the 2000 ACM conference on Computer supported cooperative work*, pp. 231–240, (2000).

6. M. Schwartz and C. Wood, Discovering shared interests using graph analysis, *Communications of the ACM.* **36**, 78 – 89 (August, 1993).
7. A. Vivacqua and H. Lieberman. Agents to assist in finding help. In *ACM Conference on Computers and Human Interface (CHI-2000)*, (2000).
8. D. Lemire, Scale and translation invariant collaborative filtering systems, *Information Retrieval.* **8**(1), 129–150, (2005).
9. B. Mirza, B. Keller, and N. Ramakrishnan, Studying recommendation algorithms by graph analysis, *Journal of Intelligent Information Systems.* **20**, 131 – 160, (March 2003).
10. J. Palau, M. Montaner, and B. Lopez. Collaboration analysis in recommender systems using social networks. In *Cooperative Information Agents VIII: 8th International Workshop, CIA 2004*, pp. 137–151, (2004).
11. A. Rashid, G. Karypis, and J. Riedl. Influence in ratings-based recommender systems: An algorithm-independent approach. In *SIAM International Conference on Data Mining*, (2005).
12. M. Balabanovic and Y. Shoham, Fab: Content-based, collaborative recommendation, *Communications of the ACM.* **40**(3), 66–72, (1997).
13. Y. Ding and X. Li. Time weight collaborative filtering. In *Proceedings of the 14th ACM International Conference on Information and Knowledge Management (CIKM '05)*, pp. 299–300, (2005).
14. J. O'Donovan and B. Smyth. Trust in recommender systems. In *Proceedings of the 10th international conference on Intelligent user interfaces*, pp. 167–174, (2005).
15. Z. Huang, W. Chung, and H. Chen, A graph model for e-commerce recommender systems, *Journal of the American Society for Information Science and Technology.* **55**(3), 259–274, (2004).
16. J. Breese, D. Heckerman, and C. Kadie. Empirical analysis of predictive algorithms for collaborative filtering. In *Proceedings of the Fourteenth Conference on Uncertainty in Artificial Intelligence.* Morgan Kaufmann, (1998).
17. K. Yu, X. Xu, J. Tao, M. Kri, and H.-P. Kriegel, Feature weighting and instance selection for collaborative filtering: An information-theoretic approach, *Knowledge and Information Systems.* **5**(2), (2003).
18. J. Herlocker, J. Konstan, A. Borchers, and J. Riedl. An algorithmic framework for performing collaborative filtering. In *SIGIR*, pp. 230–237, (1999).
19. G. Karypis. Evaluation of item-based top-n recommendation algorithms. In *CIKM*, (2001).
20. A. d. V. M. R. J. Wang. Unifying user-based and item-based collaborative filtering approaches by similarity fusion. In *SIGIR*, pp. 501–508, (2006).
21. K.-W. Cheung and L. F. Tian, Learning user similarity and rating style for collaborative recommendation, *Information Retrieval.* **7**, 395–410, (2004).
22. J. C. R. Jin and L. Si. An automatic weighting scheme for collaborative filtering. In *SIGIR*, (2004).
23. L. G. A. de Bruyn and D. Pennock. Offering collaborative-like recommendations when data is sparse: The case of attraction-weighted information filtering. In *Adaptive hypermedia and adaptive web-based systems*, (2004).
24. C. Aggarwal, J. Wolf, K.-L. Wu, and P. Yu. Horting hatches an egg: A

new graph-theoretic approach to collaborative filtering. In *Proceedings of the Fifth ACM SIGKDD Conference on Knowledge Discovery and Data Mining (KDD'99)*, pp. 201–212, (1999).
25. Z. Huang, H. Chen, and D. Zeng, Applying associative retrieval techniques to alleviate the sparsity problem in collaborative filtering, *ACM Transactions on Information Systems.* **22**(1), 116–142, (2004).
26. P. Cohen and R. Kjeldsen, Information retrieval by constrained spreading activation on semantic networks, *Information Processing and Management.* **23**(4), 255–268, (1987).
27. F. Crestani and P. Lee, Searching the web by constrained spreading activation, *Information Processing and Management.* **36**, 585–605, (2000).
28. G.-R. Xue, S. Huang, Y. Y., H.-J. Zeng, Z. Chen, and W.-Y. Ma. Optimizing web search using spreading activation on the clickthrough data. In *Proceedings of the 5th International Conference on Web Information Systems*, (2004).
29. U. Mukhopadhyay, L. Stephens, M. Huhns, and R. Bonnell, An intelligent system for document retrieval in distributed office environments, *Journal of the American Society for Information Science.* **37**(3), 123–135, (1986).
30. J. Herlocker, J. Konstan, and J. Riedl, An empirical analysis of design choices in neighbourhood-based collaborative filtering algorithms, *Information Retrieval.* **5**, 287–310, (2002).
31. Z. Huang and D. Zeng. Why does collaborative filtering work? - recommendation model validation and selection by analyzing bipartite random graphs. In *15th Annual Workshop on Information Technologies and Systems*, (2005).
32. P. Resnick, N. Iacovou, M. Suchak, P. Bergstrom, and J. Riedl. Grouplens: An open architecture for collaborative filtering of netnews. In *Proceedings of ACM 1994 Conference on CSCW*, pp. 175–186. Chapel Hill, (1994).
33. J. Herlocker. *Understanding and Improving Automated Collaborative Filtering Systems*. Phd thesis, University of Minnesota, (2000).
34. M. McLaughlin and J. Herlocker. A collaborative filtering algorithm and evaluation metric that accurately model the user experience. In *Proceedings of the 27th International ACM SIGIR Conference on Research and Development in Information Retrieval*, pp. 329 – 336, (2004).
35. D. Wu and X. Hu. Mining and analyzing the topological structure of protein-protein interaction networks. In *Symposium on Applied Computing*, pp. 185–189, (2006).
36. M. Calderon-Benavides, C. Gonzalez-Caro, J. Perez-Alcazar, J. Garcia-Diaz, and J. Delgado. A comparison of several predictive algorithms for collaborative filtering on multi-valued ratings. In *Proceedings of the 2004 ACM symposium on Applied computing*, pp. 1033–1039, (2004).

BIOGRAPHIES

Josephine Griffith graduated with a B.Sc. in computer studies and mathematics in 1994 and an M.Sc. in computing in 1997, both from the National University of Ireland, Galway. She is currently a lecturer at the National University of Ireland, Galway.

Her main research interests are in the area of collaborative filtering, more so in the specification and combination of sources of evidence in a collaborative filtering domain.

Humphrey Sorensen is a Senior Lecturer at University College Cork, Ireland, where he has worked since 1983. He was educated at University College Cork and at the State University of New York at Stony Brook. He has also taught at the University of Southern Maine and at Colby College. He teaches in the area of database and information systems.

His research has been in the area of information retrieval, filtering and visualization and in multiagent approaches to complex information tasks.

Colm O'Riordan lectures in the Department of Information Technology, National University of Ireland, Galway.

His current research focusses on cooperation and coordination in artificial life societies and multiagent systems. His main research interests are in the fields of agent based systems, artificial life and information retrieval.

PART 3
Content-based Systems, Hybrid Systems and Machine Learning Methods

Chapter 8

Personalization Strategies and Semantic Reasoning: Working in tandem in Advanced Recommender Systems

Yolanda Blanco-Fernández, José J. Pazos-Arias, Alberto Gil-Solla,
Manuel Ramos-Cabrer and Martín López-Nores

*Campus Lagoas-Marcosende, ETSE de Telecomunicacin
36310, Vigo, Spain
{yolanda, jose, agil, mramos, mlnores}@det.uvigo.es*[*]

> The generalized arrival of Digital TV will lead to a significant increase in the amount of channels and programs available to end users, making it difficult to find interesting programs among a myriad of irrelevant contents. Thus, in this field, automatic content recommenders should receive special attention in the following years to improve assistance to users. Current approaches of content recommenders have significant well-known deficiencies that hamper their wide acceptance. In this paper, a new approach for automatic content recommendation is presented that considerably reduces those deficiencies. This approach, based on the so-called Semantic Web technologies, has been implemented in the AVATAR tool, a hybrid content recommender that makes extensive use of well-known standards, such as TV-Anytime and OWL. Our proposal has been evaluated experimentally with real users, showing significant increases in the recommendation accuracy with respect to other existing approaches.

8.1. Introduction

One of the main advantages of Digital TV (DTV) is a better use of the bandwidth available for broadcasting. With the state-of-the-art compression techniques, it is usual to broadcast between four and six digital channels in the same bandwidth previously needed by an analogue one. As a result, TV spectators all around the world are beginning to access many more channels (and contents) than before, as most of the operators are

---
[*]All the authors are members of the Department of Telematics Engineering of the University of Vigo.

currently starting their digital broadcasts. This increase in the amount of available contents will be even more overwhelming after the analogue shutdown, when the spectrum used in the current analogue transmissions will be released.

Digital TV has another important strong point that permits the transmission of data and applications along with the audiovisual contents. Those applications, running on the users' receivers, are envisaged to cause a revolution in the very conception of the television. They will not only allow users to take an active role by using applications to interact with contents and service providers, but these applications will provide users with new Internet-like services, opening a new window to the Information Society for those people currently not connected to the Internet.

This new scenario, where the users will have access from their homes to a great number of contents and services from different providers, it is likely to resemble what happened with the growth of the Internet. This huge number of contents and offered services will cause the users to be disoriented: even though they may be aware of the potentiality of the system, they lack the tools to exploit it, not managing to know what contents and applications are available and how to find them. To alleviate this problem, successful search engines arose on the Internet, ranging from the syntactic matching processes of the mid-1990s to the more recent approaches to the so-called Semantic Web.[1]

Taking advantage from the experiences carried out on the Internet, both regarding the syntactic search engines and the most recent studies on the Semantic Web, this paper presents an application called AVATAR (AdVAnced Telematics search of Audiovisual contents by semantic reasoning). This is a personal assistant that assesses the adequacy of the contents offered by different providers to the preferences of every user. Its open and modular architecture is based on well-known standards in DTV domain, as we explained in Ref 2. The aim is to furnish a highly-personalized viewing experience that prevents from bewildering the users in an increasingly growing offer of TV channels.

This paper is organized as follows. Section 8.2 presents an overview of the techniques being used by the current recommenders of audiovisual contents. Section 8.3 describes the main elements of our semantic reasoning framework, including the TV ontology , the user profiles and the algorithmic details of the proposed recommendation strategy. Section 8.4 details a sample use of AVATAR. The experimental results obtained in our evaluation are shown in Sec. 8.5. Finally, Sec. 8.6 provides a discussion of the features of AVATAR, including the main conclusions from this work and motivating directions of future work.

8.2. Related Work

Currently, it is possible to identify three well-known personalization strategies in the field of recommender systems:

Content-based methods: This technique consists of suggesting to a user programs similar to those he/she liked in the past. It has been adopted in diverse real systems, such as TV Advisor^{TM},[3] TV Show Recommender,[4] and the personalized EPG proposed by Ardissono et al. in Ref. 5.

Content-based methods require a metric to quantify the similarity between the users' profiles and the target programs. To define such metric, appropriate content descriptions of the compared programs must be available, which is usually a complex and time consuming task. Several metrics have been defined in the state-of-the-art.

- Some approaches establish simple comparisons between a set of key words;
- another proposals rely on the predictive capacity of automatic classifiers (e.g. neural networks,[6] decision trees,[7] inference rules,[8-10] and Bayesian networks[4]) in order to decide the relevance of a program for a given viewer. These are computational models that classify a given input in a specific category considering a predefined training set. In the TV domain, this set stores the user's preferences (programs he/she liked in the past and their semantic attributes), the inputs are the features of a given TV program, and the output is a category that determines if this program is *appealing* or *unappealing* for the user. For that purpose, the classifiers consider the occurrence patterns of the input features in the training set.

The commented metrics have a critical weakness related to their syntactic nature, which only permits to detect similarity between items sharing the *same* attributes. For that reason, content-based recommender systems only suggest contents too similar to those known by the user, which leads to a limited diversity in the elaborated recommendations. This problem is especially serious regarding the new users of the system, as the suggestions generated for them are based on immature profiles, made up of a limited set of programs.

Collaborative filtering: This approach is based on recommending to a user those programs appealing to other like-minded viewers (named neighbors). For that purpose, this technique is focused on the ratings given by the user for each content included in his/her profile. Two different techniques have been proposed in the literature:

- User-based collaborative filtering : Two users are similar if they have rated the same items in their profiles and with similar levels of interest.
- Item-based collaborative filtering : Two items are similar if the users who have rated one of them, tend to rate the other one with similar ratings.

By its very own, collaborative filtering provides much more diverse recommendations than the content-based approaches, as it is based on the experience of the user's neighbors. In addition, collaborative filtering techniques do not require the aforementioned resource-demanding content descriptions, as they search correlations among the ratings the users assign to contents.

However, the application of the collaborative filtering on many tools, such as MovieLens,[10] TV Scout^{TM},[11] Moviefinder^{TM}, and TiVo^{TM}, revealed some drawbacks associated to this kind of approach.

- Firstly, this technique requires that some users have rated a specific content for it to be recommended. Because of this, some significant latency is observed since a new content arrives at the system till it is suggested to some user, as it is necessary that a significant number of users have previously rated it.
- Second, the lack of flexibility to estimate the similarity among users (usually based on direct overlaps among the programs in their personal profiles) leads to the so-called *sparsity problem* sparsity problem . In this case, as the number of contents in the system increases, the probability that two users have watched the same content gets lower. This reduced overlap among the users' profiles greatly hampers the discovery of similar users regarding their preferences, a critical step in collaborative approaches.
- Finally, note the so-called *gray sheep problem*, associated to those users whose preferences are "strange" — very different from the remaining viewers'. It is clear that these users have a reduced neighborhood and, therefore, receive little accurate recommendations.

Hybrid strategies: The more successful hybrid systems are those that mix content-based methods and collaborative filtering, taking the advantage of synergetic effects and mitigating the inherent deficiencies of either paradigm. This way, as Burke describes in Ref. 12, users are provided with recommendations more accurate than those offered by each strategy individually. This approach has been adopted in systems such as PTV and PTVPlus.[13]

Many existing hybrid proposals employ the so-called "*collaboration via content*" paradigm by Pazzani,[14] based on computing the similarity among users by using both the content descriptions of the products defined in their profiles (just like in content-based methods), and the levels of interest assigned to them (considered in collaborative filtering). This way, Pazzani fights the *sparsity problem* by detecting that two users have similar preferences even when there is no overlap between the products contained in their respective profiles. However, in order to measure similarity in this case, it is necessary that there exists overlap between the attributes of these products. For that reason, Pazzani's approach is still limited because of the syntactic similarity metrics used in the traditional content-based methods.

In this paper, our aim is to fight this kind of syntactic limitations by resorting to technologies borrowed from the Semantic Web field. Specifically, the main difference between the existing systems and AVATAR is the application of diverse mechanisms for representing and reasoning on the knowledge about the TV domain, taking the advantage of the inference methods developed in the Semantic Web . The goal of this reasoning process is to discover complex semantic relationships between the contents the user likes and those finally recommended to him/her. These associations, never considered in a personalization environment, go unnoticed for the existing approaches, lacking in the semantic inference capabilities supported in our approach.[15] For that purpose, our system requires (i) generic descriptions of the available TV contents, and (ii) a knowledge base on which the semantic reasoning process is applied. To this aim, AVATAR extends the metadata defined by the TV-Anytime specification,[16] and implements an ontology according to the OWL language.[17]

The use of semantic information in recommender systems has been already proposed in other systems. In the simplest proposals, the semantic descriptions are used with the goal of providing the users with additional information about the TV contents they are watching.[18,19] On the contrary, some more elaborated approaches also include these semantic attributes in the recommendation process.[20] In contrast with our approach, these proposals do not infer complex semantic relationships from the knowledge provided by the semantic descriptions. So, for example, in Ref. 20, the semantic attributes of the recommended items are considered in order to improve the offered suggestions. To this aim, the authors use a simple knowledge representation schema based on defining classes and their main semantic attributes. Taking into account this information, Ref. 20 focuses on comparing the semantic attributes of the items to the user's preferences. However, this proposal does not consider during the recommendation process more complex relationships between the classes and their instances (such as inheritance among classes and siblingness among instances or attributes). As a consequence, the approach proposed in Ref. 20 is not able to infer semantic associations like those we are interested in. This kind of associations allow to improve the quality and accuracy of the offered recommendations, as we will show in our experimental evaluation.

Our main contribution is a hybrid recommendation strategy that combines content-based methods and collaborative filtering, increasing the recommendation accuracy thanks to the aforementioned semantic inference capabilities. The cornerstone of this technique is a new and flexible metric that quantifies the *semantic similarity* between specific TV contents. Its values depend on the semantic relationships discovered between the compared programs represented in the system knowledge base.

8.3. Our Reasoning Framework

In this section, we present the main elements included in our semantic reasoning framework: the ontology about the TV domain, the user profiles, and the algorithmic details of the proposed hybrid recommendation technique.

8.3.1. *The TV ontology*

Ontologies are widely used as conceptualizations of a given application domain, where the characteristic vocabulary is identified by means of concepts and relations among them.

In the specific case of the AVATAR tool, we have implemented an ontology describing the TV domain by means of the OWL language. The core of this ontology is a class hierarchy that identifies different kinds of TV programs, being the *TV Contents* class, the more general one, on the top of the hierarchy. From it, more specific categories are defined as successive descendants until reaching the more concrete categories, known as *leaf classes*, situated in the lowest level of the hierarchy. In addition, note that we impose a tree-like structuring in this TV content hierarchy, so that each class has only one direct superclass in it. In Fig. 8.1, a reduced subset of the aforementioned TV contents hierarchy can be observed, together with the existing IS-A relationships among the classes.

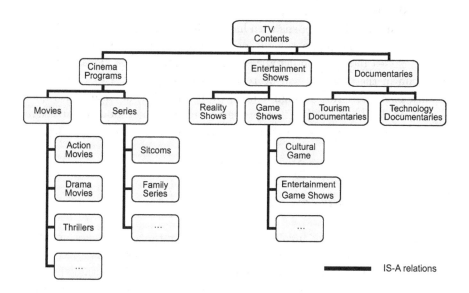

Fig. 8.1. Excerpt from our TV content hierarchy.

All the classes shown in Fig. 8.1 denote general content categories. Specific programs belonging to a given category correspond with concrete instances of the leaf classes defined in the ontology. In fact, each program can be classified into several leaf classes belonging to the hierarchy sketched in Fig. 8.1. Regarding these instances and their semantics, it is worth noting that AVATAR uses the TV-Anytime specification,[16] which describes metadata attributes for audiovisual contents such as genre, cast, topic, etc. As these attributes define the semantics of the TV programs, AVATAR automatically extracts the TV-Anytime descriptions associated to each content and translates them to the corresponding instances of the ontology. This process is simple because the classes, properties and instances of our OWL ontology have been defined according to the semantic attributes considered in the TV-Anytime specification.

Such attributes, we hereafter refer to as *semantic characteristics* of the programs, also belong to classes hierarchically organized. As some of these classes are already defined in existing ontologies, we have imported ontologies about different domains such as sports, countries, credits involved in TV contents, among others. These ontologies were extracted from the DAML repository located in the url http://www.daml.org/ontologies and converted to the OWL language by means of a tool developed by the MINDSWAP Research Group.[a]

Opposite to what happens with hierarchies, that only contain IS-A relations between concepts, ontologies permit also to define other relations between classes and between instances by means of properties. This way, each program in the ontology will be related to its respective TV-Anytime semantic characteristics through explicit properties (i.e. *hasActor*, *hasTopic*, *hasGenre*, *hasPlace*, etc.). These properties allow to infer hidden knowledge in the ontology , to be applied in the strategy proposed in Sec. 8.3.3.

8.3.2. *The User Profiles*

As AVATAR reasoning involves both the semantics of contents and the users' preferences, a formal representation of the latter is also needed to apply on them the inferential processes. Such preferences contain the programs that the user liked or disliked (positive and negative preferences, respectively) as well as those semantic characteristics relevant in a personalization environment (cast, genres, etc.). As this kind of information is already formalized in the TV ontology , our approach reuses such knowledge to model the users' preferences (for that reason the profiles used in AVATAR are named ontology-profiles).

Ontologies have already been used to model users in other works.[21] The distinctive feature of our *ontology-profiles* is that the reasoning process carried out in AVATAR requires that the users' profiles store additional semantic information. This information is added to the profiles incrementally as the system knows new data about each viewer's interests.

---

[a]http://www.mindswap.org/2002/owl.shtml.

8.3.2.1. Construction of the ontology-profiles

This progressive building of the user's ontology-profile is carried out by adding to the profile only the ontology information that identifies the specific contents associated to the positive and negative preferences of the user. When AVATAR knows a new content related to the user U, it adds to his/her profile P_U that content, the hierarchy of classes which this program belong to, and main semantic characteristics.

As an example, let us suppose that user U has watched the three following contents suggested by AVATAR: (i) the sitcom *Frasier* starring *Kelsey Grammer*, (ii) the thriller *The Silence of the Lambs*, with *Anthony Hopkins* and *Jodie Foster* in the leading roles, (iii) the game show *Wheel of Fortune*, and (iv) the technological debate *Surfing the Digital Wave*. These contents are extracted from the TV ontology by which the system knows, for example, that *Kesley Grammer* is the main star of the show *Frasier* the user U has liked. Next, these programs are translated to the profile P_U as shown in Fig. 8.2, where we can see the four instances corresponding to the watched programs, as well as the hierarchy of classes and the aforementioned semantic characteristics.

8.3.2.2. Level of interest of the users

Together with the programs and semantic characteristics of the user's preferences, the users' profiles in AVATAR also store an index to reflect the level of interest of the user in each entity. This index is known as Degree of Interest (DOI) and it is computed for the specific instances as well as for the classes contained in the ontology-profiles as described in Ref. 7. The value of the DOI index (similar to traditional explicit ratings but more complex) corresponding to a program recommended by AVATAR depends on several factors as can be the answer (acceptance or reject) of the user to the suggestion, the percentage of the program watched by the user, and the time elapsed until the user decides to watch the recommended content. The DOI of each program (always in the range $[-1, 1]$) is also used to set the DOI indexes of (i) each one of its semantic characteristics (actors, presenters, topic, etc.), and (ii) all the classes of the ontology included in the ontology-profile. In fact, the semantic characteristics of a given program inherit the DOI index of this content.

Regarding the computation of the DOI indexes for each class included in the profile, our approach firstly computes the DOI of each leaf class, and then it propagates these values through the hierarchy until reaching the

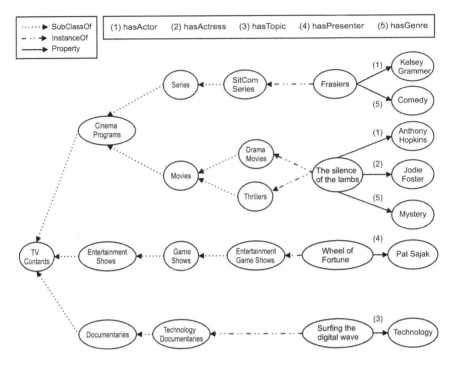

Fig. 8.2. The user ontology-profile P_U.

TV Contents class. The DOI of each leaf class is computed as the average value of the DOI indexes assigned to the programs in the profile belonging to that class. To propagate those values through the hierarchy, we adopt the approach proposed in Ref. 22 that leads to Eq. (8.1):

$$\text{DOI}(C_m) = \frac{\text{DOI}(C_{m+1})}{1 + \#\text{sib}(C_{m+1})} \quad (8.1)$$

where C_m is the superclass of C_{m+1} and $\#\text{sib}(C_m + 1)$ represents the number of siblings of the class C_{m+1} in the hierarchy of TV contents sketched in Fig. 8.1.

As a result of Eq. (8.1), this approach leads to DOI indexes higher for the superclasses closer to the leaf class whose value is being propagated, and lower for those classes closer to the root of the hierarchy (*TV Contents* class). In addition, the higher the DOI of a given class and the lower the number of its siblings, the higher the DOI index propagated to its superclass. As a class can be superclass of multiple classes, every time its DOI is updated by Eq. (8.1), our approach adds the indexes of all of its subclasses defined in the profile, and so it computes its final DOI index.

8.3.3. *A hybrid personalization technique*

Our strategy combines the content-based methods and the collaborative filtering, enhancing them with semantic inference capabilities. Its main goal is to decide to which users (named target users) a particular program must be recommended (named target content). By virtue of its hybrid nature we can identify two phases in our approach. Firstly, a content-based phase is applied, in which the approach assesses if the target content is appropriate for each target user, by considering their personal preferences. In this case, the program is suggested to these users, whereas the remaining viewers are evaluated in a second stage based on collaborative filtering.

8.3.3.1. *Content-based phase*

Given an target user U and a target content a, this phase quantifies a level of semantic matching between this content and his/her preferences defined in the ontology-profile P_U (represented as match(a, U)). The more similar the program a is to those contents most appealing to the user U, the greater the obtained semantic matching value. In order to measure this resemblance, we propose a flexible metric, named *semantic similarity*, included in Eq. (8.2):

$$\text{match}(a, U) = \frac{1}{\#N_U} \sum_{i=1}^{\#N_U} \text{SemSim}(a, c_i) \cdot \text{DOI}(c_i) \qquad (8.2)$$

where c_i is the ith content defined in the profile P_U, DOI(c_i) is the level of interest of U regarding c_i, and $\#N_U$ is the total number of programs included in P_U.

Traditional approaches just use the hierarchical structure to quantify the semantic similarity between two concepts from a taxonomy, that is, they are only based on explicit IS-A relations established in the hierarchy.[23–25] These approaches hamper the kind of complex reasoning our intelligent system requires, and for that reason, we redefine this traditional semantic similarity metric.

As far as we know, our approach is the only one that combines the IS-A relationships with the inference of more intricate ones, discovered from the properties defined in our TV ontology. Thus, to compute the semantic similarity between the target content a and a given program b, our proposal considers both the explicit knowledge represented in the TV ontology and the implicit knowledge inferred from it. The proposed semantic similarity is composed of two components corresponding to these two kinds of knowledge: hierarchical and inferential similarity. The weight of each of them is decided by a combination factor $\alpha \in [0, 1]$, as shown in Eq. (8.3).

$$\text{SemSim}(a, b) = \alpha \cdot \text{SemSim}_{\text{Inf}}(a, b) + (1 - \alpha) \cdot \text{SemSim}_{\text{Hie}}(a, b). \quad (8.3)$$

The hierarchical semantic similarity The value of the hierarchical similarity between two programs depends only on the position of the classes they belong to in the content hierarchy. In order to define its analytical expression, we use two concepts from the graph theory: depth and LCA (Lowest Common Ancestor).

The **depth** of an instance (that identifies a specific program) is equal to the number of IS-A relations between the root node of the hierarchy (*TV Contents*) and the class the instance belongs to. On the other hand, being a and b two programs, the **LCA** between them (represented as $\text{LCA}_{a,b}$) is defined as the deepest class that is ancestor of both classes a and b.

Semantic similarity is defined by Eq. (8.4):

$$\text{SemSim}_{\text{Hie}}(a, b) = \frac{\text{depth}(\text{LCA}_{a,b})}{\max(\text{depth}(a), \text{depth}(b))} \quad (8.4)$$

where:

- $\text{SemSim}_{\text{Hie}}(a, b)$ is zero if the LCA between the two programs is TV Contents (which has null depth). Otherwise, the more specific the $\text{LCA}_{a,b}$ is (i.e. deeper), the greater the similarity value.
- The closer the $\text{LCA}_{a,b}$ is to both contents' classes in the hierarchy, the higher the $\text{SemSim}_{\text{Hie}}(a, b)$ is, because the relation between a an b is more significant.

The inferential semantic similarity The inferential similarity is based on discovering implicit relations between the compared programs. These relations are inferred between those TV contents that share semantic characteristics (e.g. cast, genres, places, topics, etc.). Thus, we consider that two programs a and b are related when they are associated by properties to instances — equal or different — of a same leaf class. In case, these instances are equal, we say that the programs are associated by a *union instance*; otherwise, we say they are associated by a *union class*.

The union instances can identify any of the semantic characteristics of the compared programs. For example, two movies starring the same actor are related by the union instance that identifies him in the ontology. On the contrary, the instances of the union class can only identify some of these characteristics. Specifically, those to which a higher flexibility is allowed in the comparison (e.g. topics or locations). So, our approach can implicitly relate a World War I movie and a documentary about World War II. This association appears because both contents are related by means of the property *hasTopic* to two different instances of the union leaf class *War Topics* (*World War I* and *World War II* instances).

We define Eq. (8.5) to compute the value of $\text{SemSim}_{\text{Inf}}(a,b)$:

$$\text{SemSim}_{\text{Inf}}(a,b) = \frac{1}{\#\text{CI}_{\text{MAX}}(a,b)} \sum_{k=1}^{\#\text{CI}(a,b)} \text{DOI}(i_k) \qquad (8.5)$$

where $\#\text{CI}(a,b)$ is the number of common instances between a and b (union instances and instances of a union class), i_k is the kth of them, and $\#\text{CI}_{\text{MAX}}(a,b)$ is the maximum number of possible common instances between a and b (i.e. the minimum between the number of semantic characteristics of both programs).

According to Eq. (8.5), the higher the number of union instances and union classes between both contents, and the greater the DOI of these common instances in the user's profile we are comparing to a, the higher the value of $\text{SemSim}_{\text{Inf}}(a,b)$.

When the matching levels for all of the target users have been computed by Eq. (8.2), AVATAR suggests the target content to each user with a level over a given threshold β_{Match}. The remaining users are candidate for the collaborative phase.

8.3.3.2. *Collaborative filtering phase*

In this phase, our approach predicts the level of the interest corresponding to each candidate user U_c with respect to the target content a (represented as $\text{Pred}(U_c,a)$), based on his/her neighbors' preferences. To this aim, the existing collaborative approaches only consider those neighbors who have rated the target content. Our strategy differs from these methods because it takes into account the full neighborhood of each candidate user, during the semantic prediction process. This way, if a neighbor knows the program a, our collaborative phase uses the specific DOI he/she defined in his/her profile. Otherwise, his/her level of interest is predicted from the semantic similarity between his/her preferences and this content a.

In order to form the neighborhood of each candidate user, our approach uses the so-called *rating vectors* of users, whose components are the DOI indexes in the user's profile for the classes in the hierarchy of contents. For those classes of the hierarchy not defined in the profile, we use a zero value. After the rating vectors of each user are computed, our approach uses the Pearson-r correlation shown in Eq. (8.6) to compare them:

$$\operatorname{corr}(P_j, P_k) = \frac{\sum_r (v_j[r] - \overline{v_j})(v_k[r] - \overline{v_k})}{\sqrt{\sum_r (v_j[r] - \overline{v_j})^2 \cdot \sum_r (v_k[r] - \overline{v_k})^2}} \qquad (8.6)$$

where $\overline{v_j}$ and $\overline{v_k}$ are the mean values of the v_j and v_k rating vectors, extracted from P_j and P_k, respectively.

Notice that rating vectors do not require that two users have watched the same programs to detect if they have similar preferences. It is only necessary that the programs included in their profiles belong to the same classes in the content hierarchy. This represents a substantial reduction of the sparsity problem, typical in collaborative systems, as we said in Sec. 8.2.

Finally, the M viewers with higher correlation value with respect to the considered user form his/her neighborhood.[b] Once all of the candidate users' neighborhood has been formed, our collaborative phase computes the semantic prediction value for each one of them by applying Eq. (8.7):

$$\operatorname{Pred}(a, U_c) = \frac{1}{M} \sum_{k=1}^{M} \delta(\mathcal{N}_k) \cdot \operatorname{corr}(U_c, \mathcal{N}_k) \qquad (8.7)$$

where M is the neighborhood size, $\operatorname{corr}(U_c, \mathcal{N}_k)$ is the Pearson-r correlation between the ratings vectors of the candidate user U_c and his/her kth neighbor $\mathcal{N}_k$; and $\delta(\mathcal{N}_k)$ is a factor whose value depends on whether this neighbor has watched the program a. In case he/she does, $\delta(\mathcal{N}_k)$ is the DOI of a in $\mathcal{N}_k$'s profile; otherwise, it takes the value of the matching level $\operatorname{match}(a, \mathcal{N}_k)$ computed in the content-based phase.

According to Eq. (8.7), the value predicted to recommend a to U_c is greater when this content is very appealing to his/her neighbors, and when their respective preferences are strongly correlated. Finally, note that in this stage, we also define a threshold β_{Pred} (in $[0, 1]$) for deciding if a is recommended to each candidate user.

---

[b]The M value is chosen following heuristic criteria. In our evaluation a value between 5% and 10% of the global users has shown to lead to the best results taking into account accuracy of the recommendations and computational cost.

8.4. An Example

In this section, we describe an application scenario consisting of a specific target content and a set of target users to whom AVATAR suggests this TV program.

For the sake of clarity, some simplifications are assumed. So, we have greatly reduced both the number of semantic characteristics of each TV content, and the size of the neighborhood used in the collaborative strategy (we adopt $M = 2$). In spite of these simplifications, the shown example highlights the utility of semantic inference to compare the user's preferences, and the advantages of our method for the neighborhood formation.

In this scenario, we assume that the target content is *Kung Fu Star Search*, a reality show in which a group of people are taught by a team of kung fu masters at the Shaolin Temple, and finally, they are evaluated by a judging panel of experts, chaired by the actor *Stephen Chow*. We consider that the target users are U, N_1 and N_2, so that N_1 and N_2 are included in U's neighborhood. The programs contained in their profiles, as well as their respective DOI indexes, are shown in Table 8.1 and represented in Fig. 8.3.

Note that, according to their positive DOI indexes, all these programs were appealing to the considered users. Additionally, remember that the DOI indexes of the semantic characteristics contained in the users' profiles are the same as the indexes of the programs to which they are referred.

As shown in Table 8.1, our approach detects that the neighbors of user U are N_1 and N_2, even though none of them had watched exactly the same programs as U. Indeed, the Pearson-r correlation between U, N_1 and N_2 is high because the three viewers are interested in drama movies and sitcoms. In addition, different kinds of documentaries — *Surfing the Digital*

Table 8.1. Some TV contents defined in users' profiles.

User U	User N_1	User N_2
Wheel of Fortune (0.75)	American beauty (0.8)	Kung Fu Star Search (0.6)
Surfing the digital wave (0.9)	Welcome to Shanghai (0.85)	Mystic river (0.7)
The silence of the lambs (0.8)	The next Karate Kid (0.95)	Hannibal (0.65)
Frasier (1)	All for the winner (0.75)	Ally McBeal (0.9)

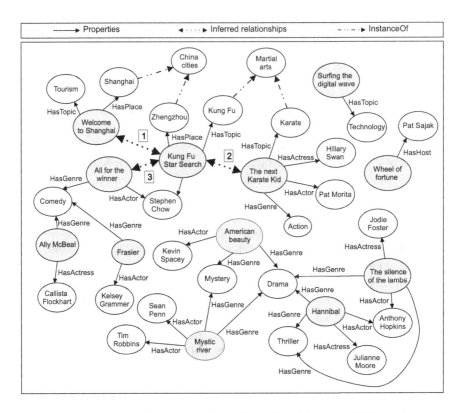

Fig. 8.3. Instances and semantic relationships inferred from our OWL ontology.

Wave and *Welcome to Shanghai* — are appealing to both U and N_1 (in Fig. 8.1, *Technology* and *Tourism documentaries*, respectively). Regarding N_2, the thrillers are interesting to both this user and U, given that they have watched *Hannibal* and *The Silence of the Lambs*, respectively.

8.4.1. A hybrid recommendation by AVATAR

In this section, we will see how the semantic inference capabilities included in AVATAR detect that *Kung Fu Star Search* is a show appealing to both the user U and his/her neighbor N_1 who have not watched it yet. Let us start summarizing in Table 8.2 the implicit semantic relationships inferred by AVATAR between the target content and the users' preferences. Next, we will describe how the discovered knowledge is used in the recommendation process.

* AVATAR discovers an implicit relation between *Welcome to Shanghai*

Table 8.2. Some semantic relationships involving *Kung Fu Star Search*.

Contents Related to *Kung Fu Star Search*	Inferential semantic similarity		Hierarchical Similarity
	Union Class	Union Instances	LCA
Welcome to Shanghai	China cities	—	TV Contents
The Next Karate Kid	Martial arts	—	TV Contents
All for the Winner	—	Stephen Chow	TV Contents
Wheel of Fortune	—	—	Entertainment Shows

and *Kung Fu Star Search* (relation 1 in Fig. 8.3), due to the fact that both contents are associated with two Chinese cities (*Shanghai* and *Zhengzhou*, respectively, instances of the join class *China cities*).[c]

∗ Even though metadata indicate that *The next Karate Kid* and *Kung Fu Star Search* involve two different martial arts, our approach infers a relationship between them by the join class *Martial Arts* (relation 2 in Fig. 8.3).

∗ As shown in Table 8.1, the movie *All for the Winner* is related to the target reality show, because *Stephen Chow* appears in both (relation 3 in Fig. 8.3).

∗ Last, *Wheel of Fortune* and *Kung Fu Star Search* are explicitly related in the TV content hierarchy sketched in Fig. 8.1, because both programs belong to the *Entertainment Shows* class (actually, these contents are instances of the classes *Entertainment Game Shows* and *Reality Shows*, respectively).

8.4.1.1. *The content-based strategy in AVATAR*

We start by applying the content-based phase on the target user U. For that purpose, it is necessary to compute the semantic similarity between the reality show *Kung Fu Star Search* and his/her preferences (see Table 8.1).

As shown in Table 8.2, only one of the programs watched by U (*Wheel of Fortune*) is related to the target reality show. Such relation is established because both contents are entertainment shows in our ontology . In this scenario, Eq. (8.4) measures a very low hierarchical similarity between these programs, because of the low depth of their LCA in the content hierarchy (in fact, depth (*Entertainment Shows*) = 1 in Fig. 8.1). Besides, Eq. (8.5)

---
[c]Note that the number of semantic characteristics in the real system is greater. So, the inferred relations are established between programs that share many common instances [see Eq. (8.5)].

leads to a null inferential similarity between the two programs due to the nonexistence of joint classes or instances between them. Combining both contributions in Eq. (8.3) results in a very low semantic matching level between U and the target reality show, whose value does not exceed the threshold β_{Match}. As a consequence, *Kung Fu Star Search* is not suggested to U, who becomes a candidate user for the collaborative phase.

8.4.1.2. *The collaborative strategy in AVATAR*

As we described in Sec. 8.2, the conventional collaborative approaches would only suggest *Kung Fu Star Search* to U when most of his/her neighbors have enjoyed this program. In our example, only the neighbor N_2 watched the reality show. Taking into account that his level of interest is moderate (0.6 in Table 8.1), the existing approaches would not suggest this program to U. Let us see how the semantic inference capabilities supported in AVATAR change this decision.

For that purpose, our collaborative phase considers all of U's neighbors, both N_2 who watched *Kung Fu Star Search*, and N_1 who does not know it. In order to predict the level of interest of N_1 in this reality show, our approach computes the semantic matching value between this program and his/her preferences [factor $\delta(\mathcal{N}_k)$ in Eq. (8.7)]. The value of *match* (N_1, *Kung Fu Star Search*) depends on the semantic similarity between the target content and those programs he/she watched in the past, as shown in Eq. (8.2). Such similarity is based on the semantic relations inferred by AVATAR that, as shown in Table 8.2, involve the most programs N_1 liked in the past. These relationships allow to discover that a reality show about Chinese martial arts will be appealing to N_1, interested in Shanghai and karate.

Noteworthy, this interest is predicted thanks to the joint classes and instances shown in Table 8.2, never considered in the traditional personalization approaches defined in the TV domain, that lack in semantic inference capabilities. In fact, the conventional content-based approaches would never select the target reality show for N_1. This is because the used similarity metrics only discovers programs belonging to the same categories to those included in his/her profile (movies and documentaries in Table 8.1), and where the same credits are involved (e.g. *Pat Morita* and *Kevin Spacey*). So, this metric leads to the well-known lack of diversity in the suggestions offered by the traditional content-based approaches, as we said in Sec. 8.2.

On the contrary, the two components of our semantic similarity allows to diversify the recommendations selected for N_1. For that purpose, our approach includes in Eq. (8.5) the DOI defined in N_1's profile for the common instances shown in Table 8.2 (DOI(*Shanghai*) = 0.85, DOI(*Kung Fu*) = 0.95 and DOI(*Stephen Chow*) = 0.75). This expression leads to a very significant inferential similarity value between *Kung Fu Star Search* and his/her preferences. This value (given that the hierarchical similarity is zero because their LCA is *TV Contents*) is used in Eq. (8.2) to compute the semantic matching level *match* (*Kung Fu Star Search*, N_1). So, our approach discovers that the reality show is a content of interest to N_1.

After the high level of interest of the neighbor N_1 with respect to *Kung Fu Star Search* has been discovered, our collaborative phase resorts to Eq. (8.7) for deciding if this reality show must (or not) be suggested to the target user U. The resulting semantic prediction value Pred (U, *Kung Fu Star Search*) is high because the two neighbors of U, strongly correlated to him/her, are interested in the target content (known interest for N_2 and predicted interest for N_1). Thanks to this relevant semantic prediction value, and opposite to what happens with existing collaborative approaches, the semantic inference supported in our strategy discovers that *Kung Fu Star Search* is a content appealing to U (and to his/her neighbor N_1).

8.5. Experimental Evaluation

Our experimental evaluation was carried out on a set of 400 undergraduate students who rated (between -1 and 1) 400 specific contents extracted from our OWL ontology (positive/negative values for appealing/uninteresting programs, respectively). A brief description was included for each program, so that all of the users could even judge the contents they had never watched.

8.5.1. *Test algorithms*

The goal of our experiments were: (i) to evaluate the accuracy of the recommendations offered by the proposed hybrid strategy (henceforth *Sem-Sim*), and (ii) to compare these results with those achieved by other techniques currently available. Bearing in mind the huge amount of different recommendation strategies existing in the literature, the following criteria have been considered to select the techniques used in our evaluation.

- Due to space limitations, we have chosen three techniques that offer ac-

curate and high-quality recommendations to the end users: we will show that their accuracy results can still be improved by inferring the complex semantic relations included in our approach.
- The selected strategies are representative in the personalized recommendations field: this way, we ensure that the results achieved by our approach are compared to techniques well-known in the state-of-the-art.
- The evaluated methods share certain basis with our hybrid strategy : so, it is possible to identify easily our contributions, and also, to verify that the formalisms added to this common basis (focused on inferring semantic relationships) improve greatly the results of the currently available techniques.

Regarding this last criterion, note that the three strategies selected in our evaluation use both content-based methods and collaborative filtering , and define some kind of semantic similarity metric. However, the distinctiveness of our similarity metric lies with the discovery of complex relations between the TV contents from the semantic descriptions considered during the recommendation process.

Specifically, the first technique is a hybrid approach that mixes content-based methods and collaborative filtering , by applying association rules as similarity metric between two specific contents.[13] The second strategy uses an approach of *pure* item-based collaborative filtering.[26] Finally, the third evaluated technique extends this *pure* item-based collaborative filtering by adding semantic information during the recommendation process.[20] However, as previously noted, this approach does not infer from this information semantic relationships like those proposed in our approach.

8.5.1.1. *Approach based on association rules (Asso-Rules)*

It is a hybrid approach that mixes a item-based collaborative phase with a content-based stage to select personalized recommendations for a set of users.[13] The collaborative phase compares (content to content) the preferences of each target user to the profiles of the remaining users. Then, it extracts his/her k most similar neighbors.

The similarity between contents is based on identifying association rules between them. These rules are automatically extracted by the Apriori algorithm,[27] which works on a set of training profiles containing several TV contents. In this approach, opposite to what happens with our semantic associations, the discovery of the association rules between two programs does not depend on their semantics, but only on their occurrence frequency in the training profiles. In fact, A and B being two contents, the rule $A \Rightarrow B$ means that if a profile contains A, it is likely to contain B as well. In fact, each rule $A \Rightarrow B$ has a confidence value interpretable as the conditional probability $p(B/A)$. This way, O'Sullivan et al.[13] compute the similarity between A and B as the confidence of the rule that involves both contents, as long as this value is greater than a threshold confidence (set in Apriori).

After the k profiles most similar to the target user have been identified, the approach *Asso-Rules* applies the content-based phase. Here, the programs contained in the profiles of these neighbors (and unknown for the target user) are ranked according to their relevance, and the top-N are returned for recommendation. A program is more relevant when (i) it is very similar to those contained in the profile of target user, (ii) it occurs in many neighbors' profiles, and (iii) these neighbors' profiles are strongly correlated to the preferences of the target user. In our evaluation, instead of selecting N contents, we used a threshold $\delta_{\text{asso-rules}}$, suggesting to the target user those programs whose relevance exceeded this threshold.

8.5.1.2. *Item-based collaborative filtering approach (Item-CF)*

This approach works on the profiles of the target users (containing a list of items and their ratings) and on a set of generic target items. We apply this approach to the domain of the personalized TV by considering that these items are TV contents. Given a target content t, the goal is to predict the level of interest of a target user U with respect to it. For that purpose, firstly, Ref. 21 must select the N contents included in U's profile that are most similar to the target program t. To compute the similarity between two programs, Ref. 26 selects those users who have rated both programs in their profiles and it computes the Pearson-r correlation between their ratings (ratings for each program and average ratings are used). So, two items are similar when many users have rated them simultaneously and with analogous ratings.

Finally, *Item-CF* predicts the interest of U in t as the weighted average of the ratings assigned by U to the N programs most similar to the target content. These weights are the similarity measures (between these N programs and t) computed by the aforementioned Pearson-r correlation. So, the more similar t to the programs appealing to U, the greater his/her level of interest in this content. In our evaluation, this predicted level of interest is compared to a threshold $\delta_{\text{item-cf}}$ for deciding if the target content is suggested to the target user (like in Asso-Rules approach).

8.5.1.3. *Semantically enhanced item-based collaborative filtering (Sem-ItemCF)*

Like in the Item-CF approach, the goal is to predict the rating of the target user U on the target item (TV content) t. With the exception of the metric of similarity between items, the *Sem-ItemCF* approach[20] is identical to *Item-CF*: both proposals select the N contents in U's profile which are most similar to the target program t. Next, these works predict the level of interest of the user U on t. This predicted level is computed as the sum of the ratings given by U on the N contents more similar to t. Each rating is weighted by the corresponding similarity between t and N selected programs.

As shown in Ref. 26, in the *pure* item-based collaborative filtering proposed in *Item-CF*, the similarity between two items i_p and i_q is computed by considering only the ratings of those users who have rated both items in their profiles (and by applying on them the expression of the Pearson-r correlation as mentioned in Sec. 8.5.1.2). In contrast, in the *enhanced* item-based collaborative described in *Sem-ItemCF*, the authors combine this component [represented as RateSim(i_p, i_q) in Eq. (8.8)] with other contributions that consider the semantic attributes of the compared items [SemSim(i_p, i_q) in Eq. (8.8)]. As shown in this expression, this approach combines linearly both components by a parameter α.

$$\text{CombinedSim}(i_p, i_q) = \alpha \cdot \text{SemSim}(i_p, i_q) + (1 - \alpha) \cdot \text{RateSim}(i_p, i_q). \quad (8.8)$$

In order to compute the semantic similarity in Eq. (8.8), the authors build a matrix S where each row refers to each one of the n items available in the recommendation process, and each column corresponds to the values of their respective semantic attributes. Once this matrix has been created, the Sem-ItemCF approach applies techniques based on Latent Semantic Indexing (LSI) to reduce its dimension.[28] This results in a much less sparse matrix S', improving the computational costs associated with the process. By virtue of the reduction, in the matrix S' each item is represented by a group of latent variables, instead of the original attributes. These variables refer to sets of highly correlated semantic attributes in the original data. According to the mathematical procedure detailed in Ref. 20, the authors obtain from S' a $n \times n$ square matrix in which an entry i, j corresponds to the semantic similarity between the items i and j (that is, SemSim(i, j)).

Note that in our application domain, the rows of the matrix S refer to the TV contents available in the experimental evaluation, and its columns are the values of their semantic characteristics, represented in our OWL ontology. Once the level of interest of U on each target content t has been computed, our experiment compares this predicted rating to a threshold $\delta_{\text{sem-itemcf}}$ for deciding if the target content is suggested to the target user (like in the previous approaches).

8.5.2. Test data

The 400 users' preferences were divided in two groups:

Training users (40%): The programs rated by these users with positive DOI indexes were used to build the so-called training profiles. From these profiles, the similarity between two specific contents is computed both in the *Asso-Rules*, the *Item-CF*, and the *Sem-ItemCF* approaches:

- Regarding the rules-based work,[13] these profiles are used to train the Apriori algorithm, whose task is to discover association rules between the programs contained in them and to quantify their similarity.
- With respect to *Item-CF*,[26] remember that the similarity between two programs is measured by selecting those training users interested in both of them, and computing the correlation Pearson-r between their respective ratings.
- The same procedure is applied in the Sem-ItemCF approach[20] to compute the component RateSim in Eq. (8.8). On the contrary, the semantic similarity used in this expression is obtained by considering the semantic characteristics represented in the OWL ontology for the 400 TV contents considered in our evaluation and building the matrix S defined in Sec. 8.5.1.3. Next, the methods described in Ref. 20 are applied on S in order to compute the square matrix of similarity between items.

Test users (60%): The four evaluated approaches are applied on their preferences. Out of the initially rated 400 programs, we selected for each user the 10 programs most appealing to him/her and the 10 he/she found less interesting. With these positive and negative preferences, their ontology-profiles were built according to as mentioned in Sec. 8.3.2. In this process, a low overlap between these users' profiles was obtained, leading a great sparsity level (89%). The remaining 380 programs rated by these users and their DOI indexes (named hereafter *evaluation data*) were hidden in order to be compared to the programs offered by each evaluated approach. This way, each test user had a profile (with 20 contents) on which each evaluated strategy was applied, and some evaluation data (380 contents) with which the finally suggested programs were compared.

8.5.3. *Methodology and accuracy metrics*

First, we run the four evaluated strategies (*Asso-Rules, Item-CF, Sem-ItemCF* and *Sem-Sim*) with the profiles of the test users as input. In addition to these profiles, *Item-CF, Sem-ItemCF* and our approach need a set of target contents, and the 400 programs the users have rated were used to fulfill this role. Besides, *Sem-ItemCF* also requires to build the matrix S defined in Sec. 8.5.1.3 to compute the semantic similarity between two TV contents. Its rows are the 400 TV contents available in our experiment, and its columns refer to the values of their semantic characteristics, represented in our OWL ontology. Regarding the *Asso-Rules* approach, the neighborhood of each test user is built — computing the similarity among programs by the association rules — and the more relevant contents are selected for each test user (excluding those ones already known by him/her).

With regard to the configuration parameters, we have used: $k = 30$ neighbors and $\delta_{\text{asso-rules}} = 0.6$ for the *Asso-Rules* approach; $N = 10$ programs and $\delta_{\text{item-cf}} = 0.6$ for *Item-CF*; $\delta_{\text{sem-itemcf}} = 0.6$ and $\alpha = 0.65$ for Sem-ItemCF, and $M = 30$ neighbors and $\beta_{\text{Match}} = \beta_{\text{Pred}} = 0.6$ for our hybrid strategy.[d]

We computed for each user the values of the precision metrics used in our evaluation: recall and precision. Recall is the percentage of programs rated as interesting by the user (that is, having a positive DOI in his/her evaluation data) that are suggested by the strategy. Precision is the percentage of the suggested contents that have a positive DOI in the user evaluation data.

Finally, we computed the average and variance of recall and precision, considering the 380 evaluated users (represented in Fig. 8.4), to be able to detect sharp dispersions between the average value and the one of each user.

8.5.4. *Assessment of experimental results*

The results of the evaluation show that the semantic inference capabilities of our strategy achieve a significant increase in the quality of the offered recommendations, even when the sparsity level of the used data is very

---
[d]After several tests, these values were observed to provide the more precise recommendations for each evaluated strategy.

high. This conclusion can be drawn from the average values of the recall and precision of the four evaluated approaches, shown in Fig. 8.4. From these values, it is clear that our approach is the one that discovers a greater number of programs appealing to the user (a higher recall) and a small number of programs not appealing (higher precision).

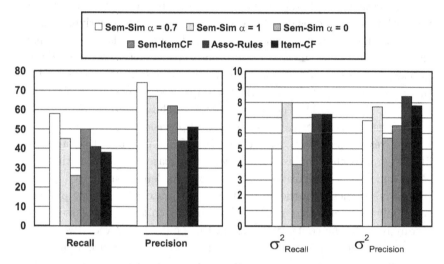

Fig. 8.4. Average (left) and variance (right) of recall and precision for the evaluated approaches.

This significant improvement is related to the similarity metric used in each evaluated approach. The implicit relationships considered in our inferential similarity are discovered thanks to the precise knowledge stored in the TV ontology. On the contrary, the rules discovered in the *Asso-Rules* approach depend on the patterns found by the *Apriori* algorithm in the training profiles, which leads to many semantic associations going unnoticed (those that cannot be derived from the identified patterns). Something similar happens to the *Item-CF* approach, where the similarity between two programs only depends on the number of users that have rated both contents in the training profiles, and on their respective ratings.

As shown in Fig. 8.4, the *enhanced* item-based collaborative filtering proposed in *Sem-ItemCF* is able to increase greatly both the recall and precision of the recommendations with respect to the *Asso-Rules* and *Item-CF* approaches. *Sem-ItemCF* overcomes the limitations of the similarity metrics used in both approaches by combining the users' ratings contained in the training profiles with the semantic attributes of the compared contents [see Eq. (8.8)]. However, the results of *Sem-ItemCF* are not so good as those achieved in our hybrid approach when both components of the semantic similarity are considered (*SemSim* with $\alpha = 0.7$ in Fig. 8.4). This is because *Sem-ItemCF* only uses part of the knowledge available in our OWL ontology in order to measure the semantic similarity. This knowledge consists of the semantic characteristics of the compared contents. Nevertheless, the authors do not consider the whole structure of the underlying domain ontology, and as a consequence, they do omit complex relationships among classes and instances inferred in our approach. In fact, *Sem-ItemCF* takes into account the union instances used in our proposal, but it obviates both the union classes and the IS-A relationships used in our hierarchical similarity (see Sec. 8.3.3.1). The absence of these relationships in *Sem-ItemCF* is the cause of its lower values of recall and precision.

The results also show the advantages of our user-based collaborative filtering approach compared to the *pure* item-based versions of the *Asso-Rules* and *Item-CF* algorithms. Our approach detects similarity between two users' profiles even though they have not watched the same programs (it is only necessary that such contents belong to the same classes in the hierarchy). On the contrary, to detect the similarity between two specific items, *Asso-Rules* and *Item-CF* only consider their occurrence frequency in training user's profiles. So, it is not possible to detect that two programs are similar if they do not appear together in such profiles. This drawback is overcome in *Sem-ItemCF* thanks to the attributes matrix S from which the semantic similarity between two contents is computed, regardless of their appearance frequency in the training profiles. Even though the semantic attributes considered in this approach improve the recommendations offered by *Asso-Rules* and *Item-CF*, this approach is less accurate than ours, as shown in Fig. 8.4. This is related to the absence of the complex semantic relationships inferred in our approach from the TV ontology, as we mentioned before.

To validate the average values of recall and precision, shown in Fig. 8.4, we also computed their variances. The low levels obtained in both proposals reveal a limited dispersion between the values of both metrics and their average values. This is a clear evidence of the representativeness of the averages our evaluation is based on.

Last, regarding α parameter in Eq. (8.3), highlight that the best results have been obtained when both components of our semantic similarity (hierarchical and inferential) are considered, and when the inferential component is favored, as shown in Fig. 8.4 for $\alpha = 0.7$. Concretely, the low recall and precision values for $\alpha = 0$ stress the benefits of the semantic inference. This time, when the discovery of new knowledge from the OWL ontology is omitted and only the IS-A relationships of Fig. 8.1 are considered, the recommendations are less precise than when $\alpha \neq 0$.

8.6. Final Discussion

TV is currently facing in the near future, plenty of significant changes and challenges, mainly derived from the introduction of software and computer-like technologies. In addition, the digital switchover will make available to the users many more channels and contents than today. As a result, assistance mechanisms will be needed to help the users to find appealing contents, avoiding interesting programs from going unnoticed.

Nowadays, significant drawbacks have been identified in the existing automatic recommenders. In this paper, we have presented a hybrid recommendation strategy for a TV intelligent assistant, based on mixing the most successful personalization approaches up-to-date: content-based methods and collaborative filtering . Our approach complements these personalization techniques with a new process of knowledge inference from the user's preferences and the TV content semantics. Our algorithms discover complex semantic relationships that relate the user's preferences to the finally suggested contents. Our formalization uses TV standard technologies as TV-Anytime, and is based on data structures (ontologies) and languages (OWL) extracted from the Semantic Web field to build the inference framework.

The proposed hybrid strategy greatly reduces the sparsity problem of the collaborative filtering approaches, thanks to a technique that uses a hierarchy to identify the general categories which the user's preferences belong to, instead of identifying specific programs. This way, we exploit this hierarchical structure in order to generate overlap between profiles containing different TV contents. In addition, this technique also alleviates the latency problem of the collaborative systems. This limitation is especially critical in the TV domain, where many contents appear continuously. In AVATAR it is possible to reduce its negative effects, by providing the viewers with the newest programs with no delay.

To alleviate the lack of diversity usually associated to the content-based methods , we have redefined the traditional semantic similarity notion. It is a new flexible metric to compare TV contents based on inferring complex semantic relationships between programs, beyond IS-A relations considered in other existing approaches. These relationships are discovered between programs that share semantic characteristics relevant to the users (e.g. cast, genres, etc.). So, the greater the number of common semantic characteristics, the higher the measured similarity.

Finally, we have shown an example of the reasoning carried out in AVATAR, highlighting appealing recommendations offered by the new strategy that could not be discovered by traditional approaches. Also, the results from our experimental evaluation revealed that the semantic inference represents a step forward in the future for personalized TV.

Our main future work is to continue the experimental evaluation of the presented hybrid strategy . Even though the first results are encouraging, more comparisons with current approaches must be undertaken for final conclusions to be drawn.

Acknowledgements

Authors thank to Jorge Garca-Duque, Rebeca P. Daz-Redondo, Ana Fernndez-Vilas and Jess Bermejo-Muoz for their collaboration and support in this work. Also, authors are grateful to the Spanish Ministry of Education and Science for funding by Research Project TSI2007-61599.

References

1. G. Antoniou and F. van Harmelen, *A Semantic Web Primer*. The MIT Press, (2004).
2. Y. Blanco, J. Pazos, M. Lpez, A. Gil and M. Ramos, AVATAR: An improved solution for personalized TV based on semantic inference, *IEEE Transactions on Consumer Electronics*. **52**(1), 223–232, (2006).
3. D. Das and H. ter Horst. Recommender systems for TV. In *Recommender Systems: Papers from the AAAI Workshop. Technical Report WS-98-08*, pp. 151 – 160. American Association for Artificial Intelligence, Menlo Park, California, (2001).
4. J. Zimmerman, K. Kurapati, A. Buczak, D. Schafer, S. Gutta and J. Martino. TV Personalization System: Design of a TV Show Recommender Engine and Interface. In *Personalized Digital Television: Targeting Programs to Individual Viewers*, pp. 27 – 51. Kluwer Academics Publishers, (2004).
5. L. Ardissono, C. Gena, P. Torasso, F. Bellifemine, A. Difino and B. Negro. User modeling and recommendation techniques for personalized electronic program guides. In *Personalized Digital Television: Targeting Programs to Individual Viewers*, pp. 3 – 26. Kluwer Academics Publishers, (2004).

6. G. Boone. Concept features in RE:Agent, an intelligent email agent. In *Proceedings of the 2nd International Conference on Autonomous Agents (Agents-98)*, pp. 141–148, (1998).
7. M. Pazzani, J. Muramatsu and D. Billsus. Syskill and Webert: Identifying interesting Web sites. In *Proceedings of the 13th National Conference in AI (AAAI-96)*, pp. 54–61, (1996).
8. B. Smyth and P. Cotter. Surfing the digital wave: generating personalized tv listings using collaborative, case-based recommendation. In *Proc. of the 3rd International Conference on Case-Based Reasoning (CBR-99)*, pp. 561 – 571, (1999).
9. C. Basu, H. Hirsh and W. Cohen. Recommendation as classification: using social and content-based information in recommendation. In *Proceedings of the 15th National Conference on Artificial Intelligence (AAAI-98)*, pp. 714–720, (1998).
10. B. Miller, I. Albert, S. Lam, J. Konstan and J. Riedl. Movielens unplugged: experiences with an occasionally connected recommender system. In *Proc. of the 8th ACM Conference on Intelligent User Interfaces (IUI-03)*, pp. 263 – 266, (2003).
11. P. Baudisch and L. Bruekner. TV Scout: guiding users from printed TV program guides to personalized TV recommendation. In *Proc. of the 1st Workshop on Personalization in Future TV (TV-01)*, pp. 151 – 160, (2001).
12. R. Burke, Hybrid Recommender Systems: Survey and Experiments, *User Modeling and User-Adapted Interaction.* **12**(4), 331–370, (2002).
13. D. O'Sullivan, B. Smyth, D. Wilson and K. McDonald, Improving the Quality of the Personalized EPG, *User Modeling and User-Adapted Interaction.* **14**, 5–36, (2004).
14. M. Pazzani, A framework for collaborative, content-based and demographic filtering, *Artificial Intelligence Review.* **13**(5), 393–408, (1999).
15. Y. Blanco, J. Pazos, A. Gil and M. Ramos. AVATAR: A Flexible Approach to Improve the Personalized TV by Semantic Inference. In *Proc. of Workshop on Web Personalization, Recommender Systems and Intelligent User Interfaces (WPRSIUI-05)*, (2005).
16. TV-Anytime Specification Series S-3 on Metadata. (2001).
17. D. McGuinness and F. van Harmelen, OWL Web Ontology Language Overview. W3C Recommendation. (2004).
18. N. Dimitrova, J. Zimmerman, A. Janevski, L. Agnihotri, N. Haas and R. Bolle. Content augmentation aspects of personalized entertainment experience. In *Proc. of 3rd Workshop on Personalization in Future TV (TV-03)*, pp. 42–51, (2003).
19. A. Prata, N. Guimaraes and P. Kommers. iTV enhanced systems for generating multi-device personalized online learning environments. In *Proc. of 4th Workshop on Personalization in Future TV (TV-04)*, pp. 274–280, (2004).
20. B. Mobasher, X. Jin and Y. Zhou. Semantically Enhanced Collaborative Filtering on the Web. In *Web Mining: applications and techniques.* IDEA Group Publishing, (2004).
21. S. Middleton. *Capturing Knowledge of User Preferences with Recommender*

Systems. PhD thesis, University of Southampton, (2003).
22. C. Ziegler, L. Schmidt-Thieme and G. Lausen. Exploting Semantic Product Descriptions for Recommender Systems. In *Proc. of 2nd ACM SIGIR Semantic Web and Information Retrieval Workshop (SWIR-04)*, (2004).
23. P. Ganesan, H. Garcia and J. Widom, Exploiting Hierarchical Domain Structure to Compute Similarity, *ACM Transactions on Information Systems.* **21** (1), 64–93, (2003).
24. D. Lin. An Information-Theoretic Definition of Similarity. In *Proc. of 15th International Conference on Machine Learning (ICML-98)*, pp. 296–304, (1998).
25. P. Resnik, Semantic Similarity in a Taxonomy: An Information-Based Measure and its Application to Problems of Ambiguity in Natural Language, *Journal of Artificial Intelligence Research.* **11**, 95–130, (1999). URL citeseer.ist.psu.edu/resnik99semantic.html.
26. B. Sarwar, G. Karypis, J. Konstan and J. Riedl. Item-based collaborative filtering recommendation algorithms. In *Proc. of 10th International World Wide Web Conference (WWW-01)*, pp. 285–295, (2001).
27. R. Agrawal and R. Srikant. Fast Algorithms for Mining Association Rules. In *Proc. of 20th Conference on Very Large Data Bases (VLDB-94)*, pp. 487–499, (1994). URL citeseer.ist.psu.edu.
28. M.W. Berry, S.T. Dumais and G.W O Brien, Using linear algebra for intelligent information retrieval, *SIAM Review.* **37**(1), 573–595, (1994).

BIOGRAPHIES

Yolanda Blanco an assistant teacher in the Department of Telematics Engineering at the University of Vigo, where she received her Ph.D. in computer science in 2007. She is a member of the Interactive Digital Lab and she is currently engaged in services for Interactive Digital TV.

José J. Pazos received his Ph.D. degree in computer science from the Department of Telematics Engineering, University of Vigo in 1995. Currently, he is involved in middleware and applications for Interactive Digital TV.

Alberto Gil received the Ph.D. degree in computer science from the University of Vigo in 2000. Currently, he is an associate professor in the Department of Telematics Engineering at the same University. He works on middleware design and interactive multimedia services.

Manuel Ramos received his Ph.D. degree in computer science from the University of Vigo in 2000. Currently, he is an associate professor in the Department of Telematics Engineering at the same University. He is engaged with personalization services for Digital TV and integration with smart home environments.

 **Martín López** an associate professor in the Department of Telematics Engineering at the University of Vigo, where he received her Ph.D. in computer science in 2006. Currently, he is involved in the design and development of Digital TV services.

Chapter 9

Content Classification and Recommendation Techniques for Viewing Electronic Programming Guide on a Portable Device

Jingbo Zhu*, Matthew Y. Ma†, Jinhong K. Guo‡ and Zhenxing Wang⁺

*Institute of Computer Software and Theory, Northeastern University
Shenyang, P.R. China
*⁺zhujingbo@mail.neu.edu.cn
⁺wzhx1983@gmail.com

†Scientific Works, USA
†mattma@ieee.org

‡Panasonic Princeton Laboratory, USA
‡kguo@research.panasonic.com

With the merge of digital television (DTV) and exponential growth of broadcasting network, an overwhelmingly amount of information has been made available to a consumer's home. Therefore, how to provide consumers with the right amount of information becomes a challenging problem. In this paper, we propose an electronic programming guide (EPG) recommender based on natural language processing techniques, more specifically, text classification. This recommender has been implemented as a service on a home network that facilitates the personalized browsing and recommendation of TV programs on a portable remote device. Evaluations of our Maximum Entropy text classifier were performed on multiple categories of TV programs, and a near 80% retrieval rate is achieved using a small set of training data.

9.1. Introduction

With the advent and widespread use of mobile phones and more recently the launch of 3G services, delivery of entertainment media and metadata such as electronic programming guide (EPG) has become more desirable. Examples of these content delivery services on a mobile network include the new DVB-H standard and the recent trials of mobile-TV service in US and Europe.

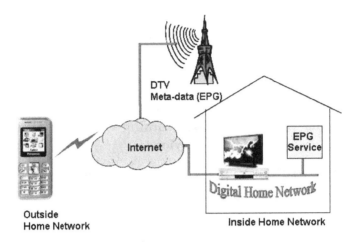

Fig. 9.1. EPG delivery system for a portable device.

However, existing technologies and infrastructures all have their shortcomings, particularly the subscription fees associated with these add-on services to the mobile network. Furthermore, only simple search/browsing functions are available due to limited computing power of a mobile phone. As the number of channels available on the TV broadcasting network increases, it becomes more challenging to deal with the overwhelmingly expanding amount of information provided by the EPG. In our view, the realization of delivering broadcast EPG information to a mobile phone must integrate a personalized EPG recommendation service that helps alleviate the burden of storage and network bandwidth requirement posted on the mobile phone. This EPG recommendation service should also leverage (a) the available content from the Internet, for which consumers have already paid premium for broadband services; (b) existing higher computing power peripherals such as digital video recorder (DVR) or media server at home; (c) open architecture and standard protocols for delivering EPG data to a mobile device. Based on these considerations, we propose an EPG delivery system for a portable device, as shown in Fig. 9.1.

In our proposed system, the EPG content comes from the Internet to leverage the existing broadband infrastructure at home. The delivery of EPG to a portable device is based on the ongoing open platform OSGi[19] and SIP.[22] The collection, processing and recommendation of EPG content are realized through an EPG Service on a home media server, which is connected to the home network. The EPG service can be realized as a software bundle that is advertized on the home network. The EPG service provides recommended EPG list to a mobile phone, as well as receives user's feedback in order to provide personalized service. The communication between EPG Service and mobile phone can be carried via SIP protocol. Details of networking architecture are described in our prior work.[5,16]

Whereas home and mobile networking are enabling technologies for the proposed EPG delivery system, the focus of this paper is on the EPG recommender system. In a home and mobile environment as illustrated in Fig. 9.1, the EPG recommender system is a crucial technology required for reducing the amount of data to be delivered to a portable device. In our research, we employ a content based recommendation engine using a statistical approach, which overcomes the disadvantages associated with traditional keyword search or keyword matching. Furthermore, we introduce the concept of "domain of interest", which truly reflects the program of interest from users' perspectives across multiple genres. The recommender based on the "domain of interest" yields higher recommendation rate than that without it.

The remainder of this paper is organized as follows. Section 9.2 describes related work. Section 9.3 introduces our proposed EPG recommender system. Our core contribution in domain identification and content recommendation are described in Sec. 9.4, followed by prototypes and experiments in Sec. 9.5. Finally, we conclude this paper in Sec. 9.6.

9.2. Related Work

Unlike the delivering of media content to portable devices, fewer efforts have been made on EPG recommendation for viewing on a portable device. In this section, we review some of prior research in the general area of EPG recommendation.

Various recommender approaches have been proposed based on the EPG content, particularly category information. Ehrmantraut *et al.*[8] and Gena[9] adopted both implicit and explicit feedback for personalized program guide. Takagi *et al.*[24] proposed a conceptual matching scheme to be applied to TV program recommendation by fusing of conceptual fuzzy sets and ontology. This work is limited to drama category and the approach is primarily based on program subcategories of drama as the top layer of the ontological structure to represent a user's taste. In a more recent approach, Blanco *et al.*[4] used TV ontology to build a collaborative recommendation system by defining similarity measures to quantify the similarity between a user's profile and the target programs. However, how to map the target program to the predefined categories is still a crucial problem, and in so-called TV ontology there is no acceptable current standard for the categories of TV programs.

Isobe et al.[11] described a set-top box based scheme that associates the degree of interest of each program with viewer's age, gender, occupation, combined with favorite program categories in sorting EPG content. Yu et al.[29] proposed an agent based system for program personalization under TV Anytime environment[25] using similarity measurement based on VSM (Vector Space Model). This work, however, assumes that the program information is available on a large storage media and does not address the problem of data sparseness and limited categories supported by most EPG providers. Pigeau et al.[20] presented a TV recommender system using fuzzy linguistic summarization technique to cope with both implicit and explicit user profiles. This system largely depends on the quality of meta-data and solely on DVB-SI standard.[23]

Cotter et al.[6] described an Internet based personalized TV program guide using an explicit profile and a collaborative approach. Xu et al.[27] also presented some interesting conceptual framework for TV recommendation system based on Internet WAP/SOAP. For portable devices, however, this system inherits the limitations of SOAP/HTTP based technologies, which add considerable network overhead on a portable device. Hatano et al.[10] proposed a content searching technique based on user metadata for EPG recommendation and filtering, in which four types of metadata are considered. The user attributes such as age, gender and place of residence are considered to implement an adaptive metadata retrieval technique. However, these attributes are too general for EPG recommendation. Personalized profiles for EPG recommendation mainly depend on users' interest or characteristics. Even though age and gender play certain roles, they are not the deciding factors.

Ardissono et al.[1,2] implemented personalized recommendation of programs by relying on the integration of heterogeneous user modeling techniques. However, users cannot often declare their preferences in a precise way.

Xu and Kenji[31] applied a simple technique for EPG recommendation in which the similarity between each new program and the pre-acquired terms from the user's watching history is estimated by inner product method, and some thresholds are used to control the number of recommendations. In real-world applications, due to EPG text content limitation, the data sparseness problem would cause serious negative effects on simple vector inner product method. And the programs whether users like or do not care are mixed in the watching history. Such situation would cause these terms acquired from the user's watching history not to exactly reflect the user's favorite programs in some senses.

Poli and Carrive[32] applied a method to model and predict TV schedules by using an extension of Markov models and regression trees in which each training example gives the start hour and day, the duration and the genre of the telecast, not involving any program text content.

Some recommender systems attempted to integrate multiple prediction methods, such as neural network, given the user's reactions to the system's recommendations. One approach to merging multiple methods is to use relevance feedback technique, which can benefit from an informed tuning parameter. For example, Shinjo et al.[21] implemented an intelligent user interface based on multimodal dialog control for audio-visual systems, in which a TV program recommender based on viewing-history analysis is included. The profiler is built using groups of keywords analyzed from EPG content.

Our work presented in this paper attempts to address two important perspectives in an EPG recommender systems: (1) a home network based framework to support the EPG recommender system for viewing on a portable device with a vision that the system is to be deployed in an embedded entertainment device in the future and targeted on a portable device (such as PDA or mobile phone) for TV program viewing; (2) a linguistic based approach to extract good feature vectors from available information to be used in a recommender classifier.

Whereas commonly used keyword based recommendation technique inherits certain limitations, we intend to focus our research on content based recommender system. The reasons are two-fold: one is that content based recommendation system provides a flexible framework that allows the integration of richer information that may truly reflect users' preferred content; secondly, EPG content available for the TV is abundant so that typical sparseness concerns associated with any content filtering and machine learning techniques are diminishing. Within the same paradigm, we also introduce a concept of "domain of interest" across multiple genres. The details of our system will be described in the following sections.

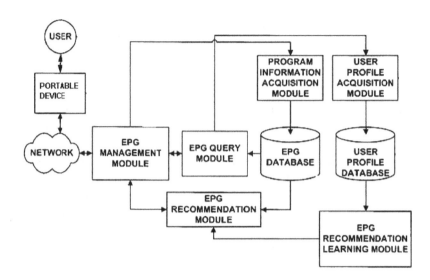

Fig. 9.2. EPG recommendation system architecture.

9.3. Proposed System

9.3.1. *Overview*

Figure 9.2 shows the architecture of our EPG recommender system. A portable device communicates with the EPG recommender system via the SIP network.[22] The EPG recommender consists of program information acquisition module, user profile module, EPG recommendation module, and EPG management and query modules.

The EPG management module is responsible for sending EPG data to and receiving user request and feedback from the portable device. Program information acquisition module collects program information from the Internet, parses text data, converts the data into a structural form usable for our recommender. Meanwhile, user profile acquisition module collects user profile data and stores it in the user profile database.

The EPG query module receives and parses the XML data in the bundle to get the content information specified by the user. The query result is packaged in the XML format, and delivered to the EPG management module in a data bundle. One copy of the query result is delivered to the user profile acquisition module for acquisition of user profile data. The user profile data primarily comprises of user's preference associated with each program. Examples of user profile data include the duration that each program is being watched (by the user), user's relevance feedback e.g. "like" or "don't care" etc.

The EPG recommendation learning module dynamically adjusts and optimizes the parameters of the recommendation algorithm according to the user profile data. The EPG recommendation module recommends programs in the database based on users' preferences. Our focus is on the recommendation module and associated learning module which will be described in detail.

9.3.2. *EPG recommender system*

Our general multi-engine EPG recommender system, as shown in Fig. 9.3, uses a series of filters to enhance the accuracy of recommendation and narrows down the search range. Five filters: time, station, category, domain, and content filter, are implemented in the recommendation process. A user can predefine a filter setting, for example, a time period from 2004-10-6::0:00 to 2004-10-8::24:00. A default time setting can also be defined, such as the current week. Time filtering can remove all programs that do not

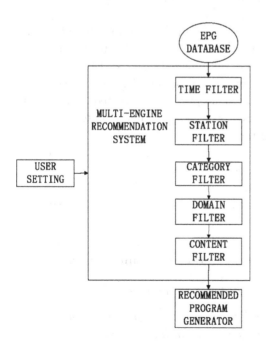

Fig. 9.3. EPG recommendation.

play within the specified time period. Station filtering works in a similar way.

Category and domain filters can be executed automatically without requiring users to preselect a criterion. Category refers to the genre of the program and it is normally available from the EPG data. Domain information more closely reflects a user's domain of interest and is broader. Domain information may cover programs across multiple categories. For example, "sports" domain includes any programs concerning sports, regardless it is a "movie", "news" or "sports" program. A trained probabilistic model is built for category filtering or domain filtering, such that the probability of a program being recommended is computed as:

$$P(c_i) = \frac{N(c_i)}{\sum_{j=1}^{|C|} N(c_j)},$$

where C denotes the set of categories/domains, c_i denotes a category/domain, and $N(c_i)$ denotes the frequency of c_i.

Since domain information is not available in provided EPG data, it can be obtained via text classification of EPG content based on the Maximum Entropy model in our system.

The content filter is designed to recommend programs based on the EPG content. It is more comprehensive as the EPG content being used in the recommendation comprises of all information in an EPG data such as station names, program titles, program descriptions, time interval, and actors. At program content level, a corpus is constructed such that it includes preferred and non-preferred programs. A binary content classifier is built using the Maximum Entropy model.

After filtering, the recommended program generator places the recommended programs into a human readable format, e.g. XML format. The formatted program information are packaged in a data bundle and sent to the portable device for display according to the user's predefined style sheet.

9.4. Domain Identification and Content Recommendation

9.4.1. *Classification problem and design choice*

A typical EPG entry has several attributes: *program title, time, channel/station, program information, duration, rating,* and *category.* The following is a sample downloaded from TV Guide[26]

Program title: *Bend It Like Beckham*
Time: *Oct 03 09:00pm*
Channel/Station: *IFC 550*
Program Info: *An 18-year-old plays for a women's soccer team but from her parents.*
Duration: *2 : 00*
Rating: *PG-13*
Category: *Movie*
Domain: Sports (need to be calculated)

As shown above, we added a new field to the downloaded EPG: Domain (of interest). As previously described, domain may reflect more of a user's interest. To perform domain filtering, domain needs to be assigned to each TV program.

In our proposed system, the classification problem is visited twice. First, program content recommendation can be formularized as a binary text classification. In other words, the task of binary classification is to automatically assign "like" or "don't care" labels to each program based on its EPG content. Secondly, the domain of interest needs to be identified for each TV program. This is a nonbinary text classification problem, in which each TV program is classified as one of the domains of interest based on its EPG content.

The text classification (TC) is a common natural language processing technique. To choose a text classifier, we considered several techniques. When provided with enough training data, a variety of techniques for supervised learning algorithms have demonstrated satisfactory performance for classification task, such as Rocchio,[12,15] SVM,[13] decision tree,[14] Maximum Entropy[18] and naive Bayes models.[17] In using these models, EPG text content can be represented as a high-dimensional vector using bag-of-words model as input for training and testing processes. All the items in the vector are treated equally. Even though feature selection is used to reduce the dimension of a vector, some salient features from diverse sources, such as TIME, STATION, ACTOR and DOMAIN, can still lose their significance in the large-dimension feature vector.

Maximum Entropy (ME) model[3] is a general-purpose machine-learning framework that has been successfully applied to a wide range of text processing tasks, such as statistical language modeling, language ambiguity resolution, and text categorization. The advantage of ME model is its ability to incorporate features from diverse sources into a single, well-balanced statistical model. In many classification tasks such as text classification[18] and spam filtering,[30] ME model often outperforms some of the other classifiers, such as naive Bayes and KNN classifier. SVM also has been performing well in many classification tasks, however, due to its high training cost, SVM is not a suitable choice for adaptive recommendation on the run. Our goal is to design a recommendation system, which dynamically updates itself as user profile is being updated. ME model is thus selected in both EPG domain identification and content recommendation.

9.4.2. *Maximum entropy model*

The domain information is identified by our system based on the original EPG content. Figure 9.4 shows a diagram of such process. Program vectors that construct the vocabulary are formed using the bag-of-words model. However, the dimension of the count matrix is very high in the feature space due to the complexity of high-dimensional text data. Therefore, feature selection is performed to lower the feature space. This step is also crucial in adapting traditional ME classifier to a mobile environment. When constructing the vocabulary, stop words are removed from the list in the training corpus.

Using bag-of-words model, a classified program can be represented as a vector of features and the frequency of the occurrence of that feature is in the form of $P = \langle tf_1, tf_2, \ldots, tf_i, \ldots, tf_n \rangle$, where $\mathbf{n}$ denotes the size of feature set, and tf_i is the frequency of the ith feature. Given a set of training samples $\mathbf{T} = \{(\mathbf{x_1}, \mathbf{y_1}), (\mathbf{x_2}, \mathbf{y_2}), \ldots, (\mathbf{x_N}, \mathbf{y_N})\}$ where $\mathbf{x_i}$ is a real value feature vector and $\mathbf{y_i}$ is the target domain, the maximum entropy principle states that data $\mathbf{T}$ should be summarized with a model that is maximally noncommittal with respect to the missing information. Among distributions consistent with the constraints imposed by $\mathbf{T}$, there exists a unique model with the highest entropy in the domain of exponential models of the form:

$$P_\Lambda(y|x) = \frac{1}{Z_\Lambda(x)} \exp\left[\sum_{i=1}^{n} \lambda_i f_i(x, y)\right] \quad (9.1)$$

where $\Lambda = \{\lambda_1, \lambda_2, \ldots, \lambda_n\}$ are parameters of the model, $f_i(x, y)$'s are arbitrary feature functions of the model, and $Z_\Lambda(x) = \sum_y \exp[\sum_{i=1}^{n} \lambda_i f_i(x, y)]$ is the normalization factor to ensure $P_\Lambda(y|x)$ is a probability distribution. Furthermore, it has been shown that the ME model is also the Maximum

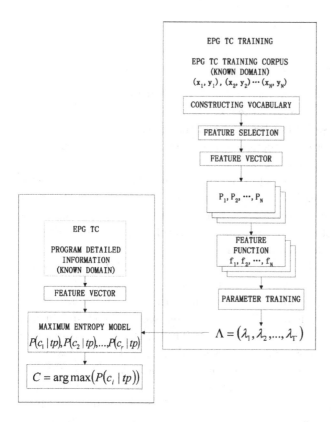

Fig. 9.4. ME-based classification process.

Likelihood solution on the training data that minimizes the Kullback–Leibler divergence between P_Λ and the uniform model. Since the log-likelihood of $P_\Lambda(y|x)$ on training data is concave in the model's parameter space Λ, a unique maximum entropy solution is guaranteed and can be found by maximizing the log-likelihood function:

$$L_\Lambda = \sum_{x,y} \tilde{p}(x,y) \log p(y|x)$$

where $\tilde{p}(x,y)$ is an empirical probability distribution. Our current implementation uses the Limited-Memory Variable Metric method, called L-BFGS, to find Λ. Applying L-BFGS requires evaluating the gradient of the object function L in each iteration, which can be computed as:

$$\frac{\partial L}{\partial \lambda_i} = E_{\tilde{p}} f_i - E_p f_i$$

where $E_{\tilde{p}} f_i$ and $E_p f_i$ denote the expectation of f_i under empirical distribution $\tilde{p}$ and model p, respectively.

9.4.3. *Feature dimension reduction*

Feature dimension reduction (also referred to as feature pruning or feature selection) is employed to reduce the size of the feature space to an acceptable level, typically several orders of magnitude smaller than the original. The benefit of dimension reduction also includes a small improvement in prediction accuracy in some cases.[28]

Two approaches, feature selection and feature extraction can be employed for this purpose. Feature selection refers to algorithms that output a subset of the input feature sets. Feature extraction creates new features based on transformations or combinations of the original feature set. Instead of using all the available features in the observation vectors, the features are selected based on some criteria of removing noninformative terms according to corpus statistics, such as document frequency, $\chi 2$ statistic, information gain, term strength and mutual information methods.[28]

The $\chi 2$ statistic is one of the best performing scoring functions for feature selection in text content classification. The $\chi 2$ statistic measures the lack of independence between a word **t** and a domain **c**. Using the two-way contingency table of a word **t** and a domain **c**, where **A** is the number of times **t** and **c** co-occur, **B** is the number of times the **t** occurs without **c**, **C** is the number of times **c** occurs without **t**, **D** is the number of times neither **c** nor **t** occurs, and **N** is the total number of training samples, the term "goodness measure" is defined to be:

$$\chi^2(t,c) = \frac{N \times (AD - CB)^2}{(A+C) \times (B+D) \times (A+B) \times (C+D)}.$$

The $\chi 2$ statistic is zero if **t** and **c** are independent. For each domain, the $\chi 2$ statistic can be computed between each entity in a training sample and that domain to extract the features. In our content recommendation system, feature selection is done by selecting words that have the highest $\chi 2$ statistic of the class variable.

9.4.4. *Domain identification*

Domain information can be obtained from a corpus of EPG data through a training process. Domain classification, formulated as a nonbinary text classification problem, is performed using Maximum Entropy classifier.

The feature function in our algorithm is defined as the following:

$$f_{w,c'}(d,c) = \begin{cases} 0 & c \neq c' \\ n(w,d) & c = c' \end{cases} \quad (9.2)$$

where $n(w,d)$ denotes the frequency of the word w in program d.

The training programs are represented as the following: $TP: tp_1, tp_2, \ldots, tp_i, \ldots, tp_n \longrightarrow T = (V, C) : (v_1, c_1), (v_2, c_2), \ldots, (v_i, c_i), \ldots, (v_n, c_n)$, where TP denotes the training program set, in which all programs are labeled as corresponding domain information. tp_i denotes training program i, V denotes the vectors, and C denotes the domains.

The feature function set F can be constructed using Eq. (2) and the parameters $\Lambda = \{\lambda_1, \lambda_2, \ldots, \lambda_n\}$ of the ME model are estimated using the feature function set F and the training samples (V, C). Using Eq. (1), given a test program $tp, P(c_1|tp), P(c_2|tp), \ldots, P(c_i|tp), \ldots, P(c_n|tp)$ for each domain can be computed. The domain $c : c = \text{argmax}(P(c_i|tp))$ is assigned to the test program tp.

9.4.5. Content classifier for recommendation

Unlike some existing systems that need users to provide keywords to establish a user profile, we utilize an explicit user feedback model. In this model, each choice by the user to indicate their preference: "like" or "don't care" is fedback into the learning module. The EPG recommendation process is also based on maximum entropy model and works in a similar way as shown in Fig. 9.4.

In EPG content recommendation, several features were extracted from the raw EPG database based on users' preference on each program. These features are divided into the following groups.

(1) Station-Name: The corresponding value for the selected station is 1.
(2) Time: Time interval the program is played. We divide a day into 24 intervals.
(3) Lexicon: Title, Episode Title, and Program Information. We construct a vocabulary using these three fields in training data. The string of the token w in Eq. (2), which is included in the vocabulary, is used as a feature.
(4) Category Feature: This is usually included in EPG data.
(5) Actors.

As shown in Fig. 9.4, feature functions are obtained from feature vectors. The calculation of the ME model Λ parameters requires the use of feature vectors and training corpus, which consists of raw EPG database and user profile. In extreme cases, if the user is only interested in one domain, the recommendation classifier would be a binary classifier that only outputs "like" or "don't care" for all program content.

9.5. Prototype and Experiments

9.5.1. *Prototype*

We built a prototype framework to enable the downloading of EPG from the Internet and viewing on a portable device. The EPG collection and recommendation system is implemented on a home network, where EPG algorithm is running on a home server that supports OSGi[19] framework. The OSGi (Open Service Gateway Initiative) framework provides an open execution environment for applications to run on heterogeneous devices. Particularly, it provides flexibility for content providers to upload updates to consumers' devices. The portable device is a Sharp Zaurus PDA with installed SIP[22] support, which allows simple text based messages to be carried between the mobile device and the home network devices.

The prototype also enables a mobile client with three functions — EPG browsing (by date, channel etc.), Program Details (for specific program) and EPG recommendation. Figure 9.5 shows a mobile user interface for (a) EPG program details and (b) a recommended program list. As shown at the bottom of Fig. 9.5(a), a "like" and "don't care" button is provided so user can give some relevance feedback to the recommendation module after reviewing the program details.

9.5.2. *Experimental database and protocol*

In our experiments, we downloaded two-weeks of EPG for 30 channels, resulting in 1Mbytes of EPG data. The training data contains 21,277 TV programs collected from DirecTV[25] between August 8 and August 12, 2005. This data is used for generating user profile data as the training corpus. The testing was performed on 7394 TV programs collected between August 13 and August 20, 2005. Each EPG data entry includes id, time, title, a brief synopsis, duration, rating and category. Most common categories contained in the EPG are series, movie, shopping, sports, special and news. The first experiment was designed to test the effectiveness of the ME classifier.

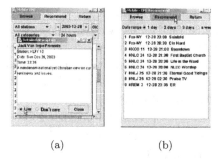

(a) (b)

Fig. 9.5. (a) EPG program details and (b) recommended program list on a mobile device.

Table 9.1. Experimental data size.

	Total Number of Programs	Training Data	Testing Data
Series	15808	11746	4062
Movie	5248	3883	1365
shopping	2576	1911	665
Sports	1807	1347	460
Special	1653	1217	436
News	1579	1173	406

The second experiment was designed to evaluate the performance of the recommender system. To do this, we divided the TV programs into groups using the category information provided by the EPG. The result of user recommendation is evaluated for each category. The corpus is collected when a user provides relevance feedback to the system upon receiving the training EPG. The testing EPG data are tagged by the same user and the recommendation result is compared with the tag.

Table 1 shows the training data and testing data sizes of the various categories of programs.

9.5.3. *Evaluation of ME classifier for domain identification*

To test the effectiveness of the text classification for domain information, we utilized the category information included in the EPG data to avoid the need to manually tag the entire database. This is because category information (tag) is readily available in EPG data whereas domain data is not. We can argue that the measuring of text classification on category can be a good indicator on the text classification performance on domain information using the same ME classifier.

For EPG training corpus, we use various fields to obtain the features of our ME model. For testing data, we remove the category information and let the ME classifier decide the category, which is then compared with the ground truth (i.e. removed category label). A quantitative measurement for performance is defined as following.

$$F1 = \frac{\text{precision} \times \text{recall} \times 2}{\text{precision} + \text{recall}}.$$

The overall F1 value is calculated as

$$\frac{\frac{1}{n}\sum_{j=1}^{n} P_j \bullet \frac{1}{n}\sum_{j=1}^{n} R_j \bullet 2}{\frac{1}{n}\sum_{j=1}^{n} P_j + \frac{1}{n}\sum_{j=1}^{n} R_j}$$

where P_j is the precision value of the jth category and R_j is the recall value of the jth category; n is the number of categories.

For our experiment, the training data includes 29,800 different words. Table 2 shows the experimental result. In this experiment, we use all EPG information as features (excluding ID and, of course, category). The F1 value is shown in Table 2.

As in Table 2, the series, movie and shopping categories yield better results. This is because there are more programs available for series, movie and shopping. Additionally, the synopsis/comments have more details for these programs. The sports and news programs, however, do not have detailed comments; some of such programs have no comments at all. The programs in special category are something difficult to classify. These programs do not fit well into all the other categories. Meanwhile, they are much diversified among themselves.

Overall, the ME classifier shows close to 0.8 F1 in the category identification, with over 0.9 on movie, followed by 0.89 F1 and 0.85 F1 for series

Table 9.2. Evaluation of ME classifier using all content.

	Precision	Recall	F1
Series	0.8680	0.9169	0.8918
Movie	0.9394	0.9137	0.9264
Shopping	0.8706	0.8283	0.8490
Sports	0.7313	0.8014	0.7647
Special	0.6968	0.5745	0.6297
News	0.8432	0.6104	0.7082
Overall F1	0.8249	0.7742	0.7987

and shopping, respectively. Although the actual performance on domain identification needs to be verified, this experiment gives promising results on the performance of our proposed ME classifier.

9.5.4. *Evaluation of content recommendation*

To conduct the experiment for content recommendation, the training and testing data were tagged by different users (four groups) as "like" and "don't care" manually. We did not use the full set of training data. Instead, we conducted experiments using training set of 100, 200, 300, 400 and 500 programs to show the effect of training data size on the recommendation result. The training data are selected randomly from the five days of training data. For example, 20 training data were selected randomly from each of the five days of training data to form the 100 training data for one experiment. A ME classifier was trained using these 100 training samples. And the testing data are processed through this classifier to get an experimental result. Such process is repeated five times using five sets of randomly selected training data. Taking 100 training data as an example, there are five randomly selected sets of training data, each containing 100 programs. Therefore, the recommendation is performed five times, and the final result for recommending the TV programs of a specific category is obtained by averaging the five testing results. Similar experiments are conducted for 200, 300, 400 and 500 training samples.

The ME classifier utilizes the following features in our experiment: time, title and program description (synopsis), program duration as well as program rating. As we have discussed above, the recommendation is performed on each category. We list the experimental results in Tables 3–8.

Table 3 shows the experiment results of TV series recommended for user A. Using 500 training samples as an example, 189 out of the 500 training samples are the ones that user selected. 1670 program segments

Table 9.3. Result of "series" recommended for user A.

Training size	100	200	300	400	500
Training data (liked)	36	74	101	145	189
Testing data (liked)	1670	1670	1670	1670	1670
Classified as liked	1295	1320	1334	1345	1360
Correctly classified	1041	1087	1102	1117	1137
Recall (%)	62.33	65.09	65.99	66.89	68.08
Precision (%)	80.39	82.35	82.61	83.05	83.60
F1	70.21	72.71	73.33	74.20	75.02

in the testing data are tagged by the user as liked series. The ME classifier classified 1360 program segments as those to recommend to the user. Out of these 1360 program segments, 1137 are correctly classified. The recall for this classifier is calculated as $1137/1670 = 0.6808$, and the precision for this classifier is calculated as $1137/1360 = 0.836$.

Similarly, we conducted experiments on movies, shopping and sports programs, as well as specials and news for user A. The results are listed in Tables 4–8, respectively.

The above results are summarized in Fig. 9.6. As shown, the horizontal axis represents the number of training samples used in training the classifier. The vertical axis represents the F1 values for each experiment. Different curves represent different categories. As training size gets larger, the F1 results get better.

Similar experiments were conducted for different users. We calculate the average F1 value of 500 training samples for different TV programs (by average) for users A–D and present this result in Fig. 9.7. As shown, the recommendation results are consistent among different users.

Further analysis on our experimental results is as follows:

1. "Movies" and "specials" cover a much broader range of interest. Thus, the synopsis is not as homogeneous as "news", "sports", etc. The F1 value is generally lower than the other categories. This is especially true for "specials" which can be so diversified that training for an efficient classifier can become a very difficult task.
2. The size of training data plays an essential role in the recommendation system. As we observe from the experimental result, the accuracy increases almost linearly with the size of the training data. In practice, the training size can be made much larger than the current experiment and

Table 9.4. Result of "movies" recommended for user A.

Training size	100	200	300	400	500
Training data (liked)	37	63	106	131	170
Testing data (liked)	427	427	427	427	427
Classified (liked)	325	353	377	384	391
Correctly classified	197	221	243	259	277
Recall (%)	46.14	51.76	56.91	60.66	64.87
Precision (%)	60.62	62.61	64.46	67.44	72.70
F1	52.26	56.63	60.45	63.98	67.64

Table 9.5. Result of "shopping" programs recommended for user A.

Training size	100	200	300	400	500
Training data (liked)	23	51	74	103	129
Testing data (liked)	165	165	165	165	165
Classified (liked)	120	136	143	147	152
Correctly classified	84	97	105	111	119
Recall (%)	50.91	58.79	63.64	67.27	72.12
Precision (%)	70.00	71.32	73.43	75.51	78.29
F1	59.23	64.47	68.19	70.90	75.02

Table 9.6. Result of "sports" programs recommended for user A.

Training size	100	200	300	400	500
Training data (liked)	43	90	132	174	221
Testing data (liked)	202	202	202	202	202
Classified (liked)	154	167	173	180	186
Correctly classified	116	127	135	143	149
Recall (%)	57.43	62.87	66.83	70.79	73.76
Precision (%)	75.32	76.05	78.03	79.44	80.11
F1	65.19	68.99	71.75	74.78	76.92

Table 9.7. Result of "specials" recommended for user A.

Training size	100	200	300	400	500
Training data (liked)	41	83	117	165	204
Testing data (liked)	163	163	163	163	163
Classified (liked)	102	113	121	127	135
Correctly classified	65	73	80	87	95
Recall (%)	39.88	44.79	49.08	53.37	58.28
Precision (%)	63.73	64.60	66.12	68.50	70.37
F1	48.83	53.10	56.44	59.88	63.65

the accuracy of the system should improve considerably. The system can also be re-trained once more data is available through user feedback.
3. The differences among different users can be attributed to the fact that some of the users' preferences happen to be the program with more ambiguous EPG description than the other user. With more training data, these differences among users should decrease.

Table 9.8. Result of "news" recommended for user A.

Training size	100	200	300	400	500
Training data (liked)	43	92	130	175	239
Testing data (liked)	184	184	184	184	184
Classified (liked)	125	143	157	164	170
Correctly classified	97	112	126	134	142
Recall (%)	52.71	60.87	68.48	72.83	77.17
Precision (%)	77.60	78.32	80.25	81.71	83.53
F1	62.57	68.54	73.79	77.04	80.06

9.5.5. *Preliminary evaluation of domain based recommendation*

Although domain based recommendation is not part of the current prototype system, we have conducted preliminary evaluation to study the behavior of recommender based on domain information. The experimental results are described as follows.

We have selected a data set of 1807 programs marked as "sports" domain by the user. These programs turned out to be distributed among multiple categories as shown in Table 9. The recommender recommends

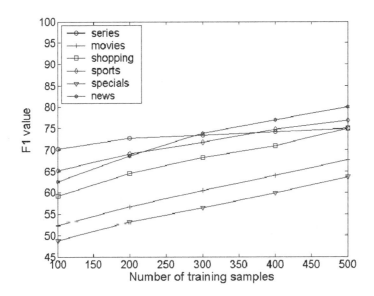

Fig. 9.6. F1 value of EPG recommendation for user A.

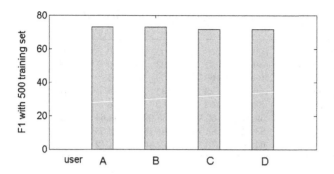

Fig. 9.7. F1 value of EPG recommendation using 500 training data.

Table 9.9. Sports domain data collected from different categories.

Series	Movie	Shopping	Sports	Special	News
408	25	310	786	71	207

"sports" related programs and this result is compared against user's predefined labels. The recommendation results are shown in Table 10.

Furthermore, the above 122 correctly classified programs from 500 training sets are distributed among multiple categories as shown in Table 11. Among the correctly classified programs for Sports domain, those categorized as "Sports" only comprise a fraction of the same domain. There are other Sports related programs in "Series", "Movie", "Shopping", "Special" and "News" categories. Clearly, the domain of interest is an influential part of recommendation, and cannot be replaced by category (genre) based recommendation. Therefore, domain based recommendation should be continuously explored in the future.

9.6. Conclusion

Among home entertainment services, electronic programming guide (EPG) is perhaps the most appealing application for television, and its services continue to grow in the emergence of the new digital TV market. Our proposed system features an EPG collection from nonproprietary data sources (i.e. HTML on the Internet) and an EPG recommender based on text classification.

Table 9.10. Experimental result of sports domain recommendation for user A.

Training size	100	200	300	400	500
Training data (liked)	39	81	123	159	208
Testing data (liked)	169	169	169	169	169
Classified (liked)	106	123	135	144	151
Correctly classified	88	103	110	116	122
Recall (%)	52.07	60.95	65.09	68.64	72.19
Precision (%)	83.01	83.74	81.48	80.56	80.79
F1	64.00	70.54	72.36	74.12	76.25

Table 9.11. Distribution of sports domain data correctly classified.

Series	Movie	Shopping	Sports	Special	News
30	3	22	53	4	10

The prototype of EPG recommender is implemented under standard open architecture home networking platform and the viewing of EPG on a portable device is enabled through existing SIP network. The presented work and prototype have suggested a feasible architecture and technology for providing a personalized EPG service that can be deployed on the home network. The average EPG browsing and recommendation time on the portable device (from user sending a UI command until the requested content is displayed) is below 1 s.

As far as we are aware, the proposed work is one of the few using natural language processing techniques for TV recommender and the result is promising. A relevance feedback is also implemented to provide dynamic personalized EPG service. The experimental results on a small scale EPG data set has shown a text classification rate of 90% and a recommendation rate of near 80%.

Our next step is to systematically collect EPG training corpus and achieve a more reliable ME model through larger scale training. Our future work also lies in further exploration of the behavior of domain information in its contribution to the recommender system. Our preliminary evaluation of domain based recommendation has already shown promising results in this perspective.

References

1. L. Ardissono, C. Gena, P. Torasso, et al., Personalized recommendation of TV programs, *Proc. 8th AI*IA Conf.*, Lecture Notes in Artificial Intelligence, Pisa (2003).
2. L. Ardissono, C. Gena and P. Torasso, User modeling and recommendation techniques for personalized electronic program guides, in *Personalized Digital Television: Targeting Programs to Individual Viewers*, Human-Computer Interaction Series, eds. L. Ardissono, M. T. Maybury and A. Kobsa, Kluwer Academic Publishers, 2004), pp. 3–6.
3. A. Berger, S. D. Pietra and V. D. Pietra, A maximum entropy approach to natural language processing, *Computer Lingustics* **22**(1) (1996) 58–59.
4. Y. Blanco, J. Pazos, M. Lopez, A. Gil and M. Ramos, AVATAR: an improved solution for personalized TV based on semantic inference, *IEEE Trans. Consumer Electronics* **52**(1) (2006) 223–232.
5. G. Chang, C. Zhu, M. Y. Ma, W. Zhu and J. Zhu, Implementing a SIP-based device communication middleware for OSGi framework with extension to wireless networks, *IEEE CS Proc. First Int. Multi-Symp. Computer and Computational Sciences (IMSCCS—06)* **2** (IEEE/Computer Society Press, China, June 2006), pp. 603–610.
6. P. Cotter and B. Smyth, PTV: intelligent personalised TV guides, *Proc. 12th Innovative Applications of Artificial Intelligence (IAAI) Conf.* (2000).
7. DirectTV Guide. http://www.directv.com
8. M. Ehrmantraut, T. Herder, H. Wittig and R. Steinmetz, The personal electronic program guide — towards the pre-selection of individual TV programs, *Proc. CIKM'96* (1996), pp. 243–250.
9. C. Gena, Designing TV viewer stereotypes for an electronic program guide, *Proc. 8th Int. Conf. User Modeling* **3** (2001), pp. 274–276.
10. J. Hatano, K. Horiguchi, M. Kawamori and K. Kawazoe, Content recommendation and filtering technology, *NTT Tech. Review* **2**(8) (2004) 63–67.
11. T. Isobe, M. Fujiwara, H. Kaneta, U. Noriyoshi and T. Morita, Development and features of a TV navigation system, *IEEE Trans. Consumer Electronics* **49**(4) (2003), pp. 1035–1042.
12. D. J. Ittner, D. D. Lewis and D. D. Ahn, Text categorization of low quality images, in *Symposium Document Analysis and Information Retrieval*, Las Vegas, 1995.
13. T. Joachims, Text categorization with support vector machines: learning with many relevant features, Machine Learning: ECML-98, *Tenth European Conference on Machine Learning* (1998), pp. 137–142.
14. D. Lewis, A comparison of two learning algorithms for text categorization, *Symposium Document Analysis and Information Retrieval* (1994).
15. D. Lewis, R. Schapire, J. Callan and R. Papka, Training algorithms for linear text classifiers, *Proc. ACM SIGIR* (1996), pp. 298–306.
16. M. Y. Ma, J. Zhu, J. K. Guo and G. Chang, Electronic programming guide recommender for viewing on a portable device, *Proc. Workshop of Web Personalization, Recommendation Systems and Intelligent User Interfaces*

(WPRSIUI) (Reading, UK, October, 2005).
17. A. McCallum and K. Nigam, A comparison of event models for naïve Bayes text classification, *AAAI-98 Workshop on Learning for Text Categorization* (1998).
18. K. Nigam, J. Lafferty and A. McCallum, Using maximum entropy for text classification, *IJCAI-99 Workshop on Machine Learning for Information Filtering* (1999), pp. 61–67.
19. OSGi: Open Services Gateway Initiative. http://www.osgi.org
20. A. Pigeau, G. Raschia, M. Gelgon, N. Mouaddib and R. Saint-Paul, A fuzzy linguistic summarization technique for TV recommender systems, *Proc. IEEE Int. Conf. Fuzzy Systems* (2003), pp. 743–748.
21. H. Shinjo, U. Yamaguchi, A. Amano, *et al.*, Intelligent user interface based on multimodel dialog control for audio-visual systems, *Hitachi Rev.* **55** (2006) 16–20.
22. SIP: Session Initiation Protocol. http://ietf.org/html.charters/sip-charter.html.
23. Specification for Service Information (SI) in DVB Systems, *DVB Document A038 Rev.* 1 (May 2000).
24. T. Takagi, S. Kasuya, M. Mukaidono and T. Yamaguchi, Conceptual matching and its applications to selection of TV programs and BGMs, *IEEE SysInt. Conf. Systems, Man and Cybernetics* **3** (1999) 269–273.
25. TV Anytime Forum. http://www.tv-anytime.org
26. TV Guide. http://www.tv-guide.com
27. J. Xu, L. Zhang, H. Lu and Y. Li, The development and prospect of personalized TV program recommendation systems, *Proc. IEEE 4th Int. Symp. Multimedia Software Engineering (MSE)* (2002).
28. Y. Yang and J. P. Pedersen, Feature selection in statistical learning of text categorization, *14th Int. Conf. Machine Learning* (1997), pp. 412–420.
29. Z. Yu, X. Zhou, X. Shi, J. Gu and A. Morel, Design, implementation, and evaluation of an agent-based adaptive program personalization system, *Proc. IEEE 5th Int. Symp. Multimedia Software Engineering (MSE)* (2003).
30. L. Zhang, J. Zhu and T. Yao, An evaluation of statistical spam filtering techniques, *ACM Trans. Asian Lang. Inform. Process. (TALIP)* **3**(4) (2004) 243–269.
31. J. Xu and A. Kenji, A personalized recommendation system for electronic program guide, *18th Australian Joint Conference on Artificial Intelligence*, (2005) pp.1146-1149.
32. J. P. Poli and J. Carrive. 2006. Improving Program Guides for Reducing TV Stream Structuring Problem to a Simple Alignment Problem. *International Conference on Computational Inteliqence for Modelling Control and Automation (CIMCA 2006)*, p.31

BIOGRAPHIES

Jingbo Zhu received a Ph.D. in computer science from Northeastern University, P.R. China in 1999, and has been with the Institute of Computer Software and Theory of the same university since then. Now he is a full professor in the Department of Computer Science, and is in charge of research activities within the Natural Language Processing Laboratory.

He has published more than 70 papers, and holds one United States patent.

Dr. Zhu's current research interests include natural language parsing, machine translation, text topic analysis, knowledge engineering, machine learning and intelligent systems.

Matthew Y. Ma received his Ph.D. from Electrical and Computer Engineering at Northeastern University, Boston, Massachusetts. He's also held MS and BS, both in Electrical Engineering from State University of New York at Buffalo and Tsinghua University, Beijing, respectively. Dr. Ma is currently with Scientific Works, USA as Chief Scientist. Prior to that, he has held several technical lead positions at government research and more recently at Panasonic R&D Company of America, where he managed a research group which focused on Panasonic's document and mobile imaging business. Dr. Ma has 11 granted US patents and is the author of 30 conference and journal publications. He is the associate editor of the International Journal of Pattern Recognition and Artificial Intelligence (IJPRAI). He is also the guest editor of IJPRAI special issues in Intelligent Mobile and Embedded Systems (2006), and Personalization Techniques and Recommender Systems (2007). Dr. Ma has been an affiliated professor at Northeastern University, China since 2002. He served as program/session chair and program committee member for numerous international conferences. His primary research interest includes image analysis, pattern recognition and natural language processing, and their applications in home networking and ambient intelligence of smart appliances.

Jinhong K. Guo received her B.S. degree from Tsinghua University in China, M.S. degree from Northeastern University in U.S. and Ph.D. degree from the University of Maryland, College Park, all in electrical engineering. She is a senior scientist with Panasonic Princeton Laboratory.

Her research interests include image and signal processing, pattern recognition, computer security, machine learning and natural language processing.

Zhenxing Wang received his B.S. degree in computer science from Northeastern University, P.R. China in 2001. He then continued his graduate study at the Institute of Computer Software and Theory in Northeastern University, Shenyang, China. Now he is a master's degree candidate supervised by Prof. Jingbo Zhu at the Natural Language Processing Laboratory.

His current research interests include natural language parsing and machine learning.

Chapter 10

User Acceptance of Knowledge-based Recommenders

Alexander Felfernig and Erich Teppan[*]
*Intelligent Systems and Business Informatics, University Klagenfurt
Universitaetsstrasse 65-67, Klagenfurt, A-9020, Austria*

Bartosz Gula [†]
*Cognitive Psychology, University Klagenfurt
Universitaetsstrasse 65-67, Klagenfurt, A-9020, Austria*

Recommender applications support decision-making processes by helping online customers to identify products more effectively. Recommendation problems have a long history as a successful application area of Artificial Intelligence (AI) and the interest in recommender applications has dramatically increased due to the demand for personalization technologies by large and successful e-Commerce environments. Knowledge-based recommender applications are especially useful for improving the accessibility of complex products such as financial services or computers. Such products demand a more profound knowledge from customers than simple products such as CDs or movies. In this paper we focus on a discussion of AI technologies needed for the development of knowledge-based recommender applications. In this context, we report experiences from commercial projects and present the results of a study which investigated key factors influencing the acceptance of knowledge-based recommender technologies by end-users.

10.1. Introduction

Recommender applications support online customers in the effective identification of products and services suiting their wishes and needs. These applications are of particular importance for increasing the accessibility of product assortments for users not having a detailed product domain knowl-

---
[*]felfernig@uni-klu.ac.at, teppan@uni-klu.ac.at
[†]gula@uni-klu.ac.at

edge. Application areas for recommender technologies range from the recommendation of financial services[1] to the personalized provision of news.[2,3] An overview of different recommender applications can be found, e.g. in Refs. 4 and 5. Compact overviews of different technological approaches to the implementation of recommender applications can be found in Refs. 6–9 and 10. There are three main approaches to the implementation of recommender applications. First, *collaborative filtering*[9,11,12] stores preferences of a large set of customers. Assuming that human preferences are correlated, recommendations given to a customer are derived from preferences of customers with similar interests. If two customers have bought similar books in the past and have rated those books in a similar way, books (with a positive rating) read by only one of them are recommended to the other. Second, *content-based filtering*[13] uses preferences of a *specific* customer to infer recommendations. In this context, products are described by keywords (categories) stored in a profile in the case that a customer buys a product. The next time, the customer interacts with the recommender application, stored preferences from previous sessions are used for offering additional products which are assigned to similar categories. Finally, *knowledge-based recommender applications* (advisors)[1,7,14,15] exploit deep knowledge about the product domain in order to determine recommendations. When selling, for example, investment portfolios, recommendations (solutions) must conform to legal regulations and suit a customer's financial restrictions as well as his/her wishes and needs. Compared to simple products such as books, movies or CDs, such products are much more in need of information and mechanisms alleviating their accessibility for customers without detailed product domain knowledge. Primarily, knowledge-based advisors provide the formalisms needed in this context.

The remainder of this paper is organized as follows. In Sec. 2, we introduce the architecture and major technologies implemented in Koba4MS,[a] a domain-independent environment designed for the development of knowledge-based recommender applications. In Sec. 3, we report experiences from successfully deployed recommender applications and discuss effects knowledge-based recommender technologies have on the behavior of customers interacting with the recommender application. Finally, in Sec. 4, we provide a discussion of related work.

---

[a]Koba4MS (*Knowledge-based Advisors for Marketing and Sales*, FFG-808479) is a research version of a commercially available recommender environment (see www.configworks.com).

10.2. Koba4MS Environment

10.2.1. Architecture

Knowledge-based advisors exploit deep product domain knowledge in order to determine solutions which suit the wishes and needs of a customer. Two basic aspects have to be considered when implementing a knowledge-based recommender application. First, the relevant product, marketing and sales knowledge has to be acquired and transformed into a formal representation, i.e. a *recommender knowledge base*[1] has to be defined. Such a knowledge base consists of a formal description of the relevant set of products, possible customer requirements and constraints defining allowed combinations of customer requirements and product properties. Second, a *recommender process*[16] has to be defined which represents personalized navigation paths through a recommender application. Both, knowledge base and recommender process design are supported on a graphical level in the Koba4MS environment. Figure 10.1 depicts the overall architecture of the Koba4MS environment. Recommender knowledge bases and process definitions are designed and maintained on a graphical level using the *development environment* (*Koba4MS Designer* and *Process Designer*) and are stored in an underlying relational database. The resulting graphical models can be automatically translated into an executable recommender application (Java Server Pages). The recommender application is made available for customers (e.g. via online-stores) and sales representatives (e.g. via intranet applications or installations on notebooks of sales representatives), where *Koba4MS Server* supports the execution of advisory sessions (*runtime environment*). Based on given user inputs, the server determines and executes a personalized dialogue flow, triggers the computation of results and determines explanations as to why a product suits the needs and wishes of a customer.

10.2.2. Recommender knowledge base

The first step when building a knowledge-based recommender application is the construction of a *recommender knowledge base* which consists of two sets of variables representing *customer properties* and *product properties* (V_C, V_{PROD}) and three sets of corresponding constraints which represent three different types of restrictions on the combination of customer requirements and products (C_C, C_F, C_{PROD}). A simplified example of a financial services knowledge base is depicted in Fig. 10.2. We will now discuss the major parts

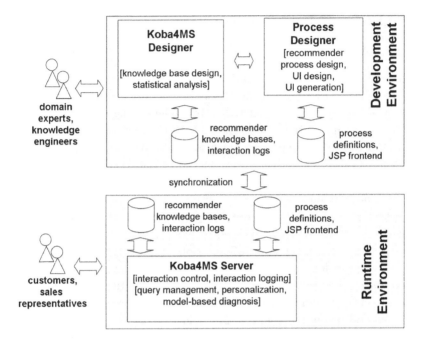

Fig. 10.1. Koba4MS architecture.

of such a knowledge base in more detail.

Customer properties: Customer properties (V_C) describe possible customer requirements related to a product assortment. *Customer requirements* are instantiations of customer properties. In the financial services domain, *willingness to take risks* ($wr_c[low, medium, high]$) is an example of such a property and $wr_c = low$ is an example of a customer requirement. Further examples of customer properties are the *intended duration of investment* ($id_c[shortterm, mediumterm, longterm]$), the *knowledge level of a customer* ($kl_c[expert, average, beginner]$), or the requested *product type* ($sl_c[savings, bonds]$) (for low risk investments).

Product properties: Product properties (V_{PROD}) are a description of the properties of a given set of products in the form of finite domain variables. Product properties in the financial services domain are, e.g. the *minimal investment period* ($mniv_p[1..14]$), the *product type* ($type_p[savings, bonds, stockfunds, singleshares]$), or the *expected return rate* ($er_p[1..40]$).

Compatibility constraints: Compatibility constraints are restricting the possible combinations of customer requirements, e.g. *if a customer has*

Customer Properties (V_C):	Compatibility Constraints(C_C):
/* level of expertise */	CC_1: $\neg(wr_c = \text{high} \wedge id_c = \text{shortterm})$
kl_c[expert, average, beginner]	CC_2: $\neg(kl_c = \text{beginner} \wedge wr_c = \text{high})$
/* willingness to take risks */	CC_3: $\neg(sl_c = \text{bonds} \wedge id_c = \text{shortterm})$
wr_c[low, medium, high]	... further compatibility constraints
/* duration of investment */	
id_c[shortterm, mediumterm, longterm]	Filter Constraints(C_F):
	CF_1: $id_c = \text{shortterm} \Rightarrow mniv_p < 3$
/* advisory wanted? */	CF_2: $id_c = \text{mediumterm} \Rightarrow mniv_p >= 3 \wedge mniv_p < 6$
aw_c[yes, no]	
/* direct product search */	CF_3: $id_c = \text{longterm} \Rightarrow mniv_p >= 6$
ds_c[savings, bonds, stockfunds, singleshares]	CF_4: $wr_c = \text{low} \Rightarrow ri_p = \text{low}$
	CF_5: $wr_c = \text{medium} \Rightarrow ri_p = \text{medium} \vee ri_p = \text{low}$
/* type of low risk investment */	
sl_c[savings, bonds]	CF_6: $wr_c = \text{high} \Rightarrow ri_p = \text{high} \vee ri_p = \text{medium} \vee ri_p = \text{low}$
/* availability of funds? */	
av_c[yes,no]	CF_7: $kl_c = \text{beginner} \Rightarrow ri_p <> \text{high}$
/* type of high risk investment */	CF_8: $sl_c = \text{savings} \Rightarrow type_p = \text{savings}$
sh_c[stockfunds, singleshares]	CF_9: $sl_c = \text{bonds} \Rightarrow type_p = \text{bonds}$
	CF_{10}: $ds_c = \text{savings} \Rightarrow type_p = \text{savings}$
Product Properties (V_{PROD}):	... further filter constraints
/* product name */	
$name_p$[text]	Allowed instantiations of
/* expected return rate */	Product Properties(C_{PROD}):
er_p[1..40]	/* product 1 */
/* risk rate of product */	CP_1: $name_p = \text{savings1} \wedge er_p = 3 \wedge ri_p = \text{low} \wedge mniv_p = 1 \wedge type_p = \text{savings} \wedge inst_p = \text{A} \vee ...$
ri_p[low, medium, high]	
/* minimum investment period */	/* product 2 */
$mniv_p$ [1..14]	CP_2: $name_p = \text{bonds2} \wedge er_p = 5 \wedge ri_p = \text{medium} \wedge mniv_p = 5 \wedge type_p = \text{bonds} \wedge inst_p = \text{B} \vee ...$
/* product type */	
$type_p$ [savings, bonds, stockfunds, singleshares]	/* product 3 */
	CP_3: $name_p = \text{stock3} \wedge er_p = 9 \wedge ri_p = \text{high} \wedge mniv_p = 10 \wedge type_p = \text{stockfunds} \wedge inst_p = \text{B}$
/* financial institute */	
$inst_p$(text)	... further product instances

Fig. 10.2. Example recommender knowledge base.

little knowledge about financial services, no high risk products should be preferred by the customer, i.e. CC_2: $\neg(kl_c = \text{beginner} \wedge wr_c = \text{high})$. Confronted with such incompatible requirements, the recommender application indicates the incompatibility and requests an adaptation of the given preferences. On the one hand, incompatibility constraints can be represented on the *textual level*. On the other hand, such constraints are represented in the form of *incompatibility tables* (tables representing not allowed combinations of customer requirements) which is the preferred representation used by domain experts designing knowledge bases (see, e.g. Ref. 1).

Filter constraints: Filter constraints (C_F) define the relationship between customer requirements and an available product assortment. Examples of filter constraints are: *customers without experiences in the financial services domain should not receive recommendations which include high risk products*, i.e. CF_7: $kl_c = beginner \Rightarrow ri_p <> high$ or *if the customer strongly prefers savings, the corresponding recommendation should include savings*, i.e. CF_{10}: $ds_c = savings \Rightarrow type_p = savings$. Figure 10.3 depicts an example of the representation of filter constraints in the Koba4MS environment (filter constraint CF_{10} of Fig. 10.2) where CF_{10} is represented on the *textual* as well as on the *graphical level*. Using the tabular representation, an arbitrary number of condition and conclusion variables can be added.

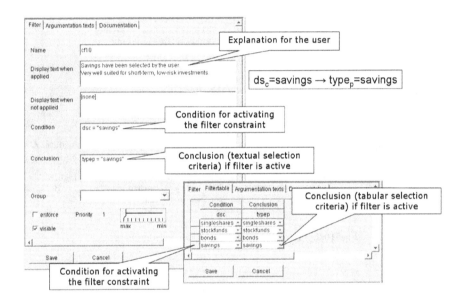

Fig. 10.3. Filter constraints: textual and graphical representation.

Product instances: Allowed instantiations of product properties can be interpreted as constraints (C_{PROD}) which define restrictions on the possible instantiations of variables in V_{PROD}, e.g. the constraint CP_2: $name_p = bonds2 \wedge er_p = 5 \wedge ri_p = medium \wedge mniv_p = 5 \wedge type_p = bonds \wedge inst_p = B$ specifies a product of type *bonds* of the financial services provider B with the name *bonds2*, an expected return rate of 5%, a *medium* risk rate, and a minimum investment period of 5 *years*.

Product comparisons: Comparison rules (see Fig. 10.4) specify which

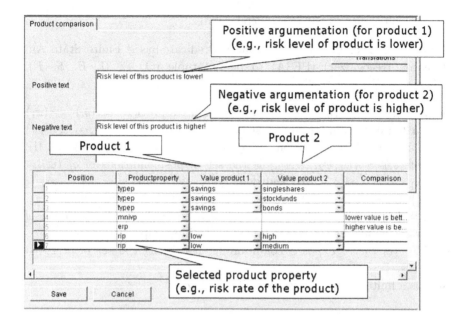

Fig. 10.4. Defining rules for product comparison.

argumentations are used to explain differences between products which are part of a recommendation result, e.g. if the risk rate of the selected product (product 1) is lower than the risk rate of another product (product 2) part of the recommendation result, then the comparison component should display the explanation *the risk level of this product is higher* (if product 1 is selected as reference product the explanation is given for all other products with a medium risk level).

10.2.3. *Recommender process definition*

In order to be able to adapt the dialog style to a customers preferences and level of product domain knowledge, we have to provide mechanisms which allow the definition of the intended (personalized) behavior of the recommender user interface . A recommender user interface can be described by a finite state automaton, where state transitions are triggered by requirements imposed by customers. Such automata are based on the formalism of *predicate-based finite state automata* (PFSA),[16] where constraints specify transition conditions between different states.

Definition 1 (PFSA). We define a Predicate-based Finite State Automaton (recognizer) (PFSA) to be a 6-tuple $(Q, \Sigma, \Pi, E, S, F)$, where

- $Q = \{q_1, q_2, \ldots, q_m\}$ is a finite set of states, where $\text{var}(q_i) = \{x_i\}$ is a finite domain variable assigned to q_i, $\text{prec}(q_i) = \{\phi_1, \phi_2, \ldots, \phi_n\}$ is the set of preconditions of q_i ($\phi_\alpha = \{c_{\alpha 1}, c_{\alpha 2}, \ldots, c_{\alpha o}\} \subseteq \Pi$), $\text{postc}(q_i) = \{\psi_1, \psi_2, \ldots, \psi_p\}$ is the set of postconditions of q_i ($\psi_\beta = \{c_{\beta 1}, c_{\beta 2}, \ldots, c_{\beta q}\} \subseteq \Pi$), and $\text{dom}(x_i) = \{x_i = d_{i1}, x_i = d_{i2}, \ldots, x_i = d_{ik}\}$ denotes the set of possible assignments of x_i, i.e. the domain of x_i.
- $\Sigma = \{x_i = d_{ij} \mid x_i \in \text{var}(q_i), (x_i = d_{ij}) \in \text{dom}(x_i)\}$ is a finite set of variable assignments (input symbols), the input alphabet.
- $\Pi = \{c_1, c_2, \ldots, c_r\}$ is a condition set restricting the set of accepted words.
- E is a finite set of transitions $\subseteq Q \times \Pi \times Q$.

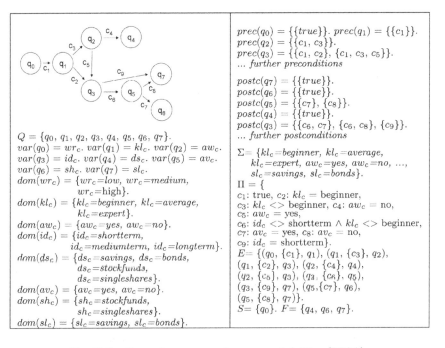

Fig. 10.5. Example recommender process definition (PFSA).

- $S \subseteq Q$ is a set of start states.
- $F \subseteq Q$ is a set of final states.

Figure 10.5 contains a PFSA definition for our example financial services recommender knowledge base depicted in Fig. 10.2. Following this definition, customers can specify requirements (input values) for the defined set of customer properties. Depending on the input of the customer, the automaton changes its state, e.g. an *expert* (c_3) *who isn't interested in financial advisory* (c_4) is forwarded to the state q_4 by the transitions $(q_0, c_1, q_1), (q_1, c_3, q_2), (q_2, c_4, q_4)$. The recommender process definition of Fig. 10.5 can be automatically translated into a corresponding recommender application. This approach allows rapid prototyping development processes for knowledge-based advisors.[16] Note that for reasons of effective knowledge acquisition support recommender process definitions are represented on a graphical level within the Koba4MS development environment (see, e.g. Ref. 16).

10.2.4. *Calculating recommendations*

We denote the task of identifying products for a customer as *recommendation task*.

Definition 2 (Recommendation Task). A recommendation task is defined by $(V_C, V_{\text{PROD}}, C_F \cup C_C \cup C_{\text{PROD}} \cup C_R)$, where V_C is a set of variables representing possible customer requirements and V_{PROD} is a set of variables describing product properties. C_{PROD} is a set of constraints describing available product instances, C_C is a set of constraints describing possible combinations of customer requirements (compatibility constraints) and C_F is a set of constraints describing the relationship between customer requirements and available products (also called filter constraints). Finally, C_R is a set of concrete customer requirements (represented as unary constraints).

Example 1 (Recommendation Task). In addition to the recommender knowledge base $(V_C, V_{\text{PROD}}, C_F \cup C_C \cup C_{\text{PROD}})$ of Fig. 10.2, $C_R = \{wr_c = low, kl_c = beginner, id_c = shortterm, sl_c = savings\}$ is a set of requirements.

Based on the given definition of a recommendation task, we can introduce the notion of a solution (*consistent recommendation*) for a recommendation task.

Definition 3 (Consistent Recommendation). An assignment of the variables in V_{PROD} is denoted as consistent recommendation for a recommendation task $(V_C, V_{\text{PROD}}, C_F \cup C_C \cup C_{\text{PROD}} \cup C_R)$ iff each variable in $V_C \cup V_{\text{PROD}}$ has an assignment which is consistent with $C_F \cup C_C \cup C_{\text{PROD}} \cup C_R$.

Example 2 (Consistent Recommendation). A consistent recommendation (result) for the recommendation task defined in Example 1 is, e.g. $name_p = savings1$, $er_p = 3$, $ri_p = low$, $mniv_p = 1$, $type_p = savings$, $inst_p = A$}.

For the calculation of solutions we have developed a *relational query-based approach*, in which a given set of customer requirements makes a conjunctive query. Such a query is composed from the consequent part of those filter constraints whose condition is consistent with the given set of customer requirements (*active* filter constraints), e.g. the consequent part of CF_7: $kl_c = beginner \Rightarrow ri_p <> high$ is translated into the expression $ri_p <> high$ as part of the corresponding conjunctive query. Accordingly, V_{PROD} is represented by a set of table attribute definitions (the product table) and C_{PROD} is represented by tuples whose values represent instantiations of the attributes defined in V_{PROD}. Furthermore, customer properties (V_C) are represented as input variables where the compatibility (C_C) of the corresponding instantiations is ensured by a consistency checker. The execution of the conjunctive query on a product table results in a set of recommendations which are presented to the customer. For the given customer requirements (C_R) of Example 1, the set $\{CF_1, CF_4, CF_7, CF_8\}$ represents active filter constraints. The consequent parts of those constraints make a conjunctive query of the form $\{mniv_p < 3 \land ri_p = low \land ri_p <> high \land type_p = savings\}$. For our example knowledge base of Fig. 10.2, this query results in the single recommendation of Example 2 $\{name_p = savings1, er_p = 3, ri_p = low, mniv_p = 1, type_p = savings, inst_p = A\}$.

Repair of Customer Requirements. If the result set of a query is empty (no solution could be found), conventional recommender applications tell the user (customer) that no solution was found, i.e. no clear explanation for the reasons for such a situation is given. Simply reporting retrieval failures (no product fulfils all requirements) without making further suggestions how to recover from such a situation is not acceptable.[17,18] Therefore, our goal is to find a set of possible compromises that are presented to the customer who can choose the most acceptable alternative. Koba4MS supports the calculation of repair actions for customer requirements (a minimal set of changes allowing the calculation of a solution). If

$C_R = \{x1_c = a_1, x2_c = a_2, \ldots, xn_c = a_n\}$ is a set of customer requirements and the recommendation task $(V_C, V_{\text{PROD}}, C_F \cup C_C \cup C_{\text{PROD}} \cup C_R)$ does not have a solution, a repair is a minimal set of changes to C_R (resulting in C'_R) s.t. $(V_C, V_{\text{PROD}}, C_F \cup C_C \cup C_{\text{PROD}} \cup C'_R)$ has a solution. The computation of repair actions is based on the Hitting Set algorithm[19,20] which exploits minimal conflict sets[21] in order to determine minimal diagnoses and corresponding repair actions.

A simple example of the calculation of repair actions is depicted in Fig. 10.6. In this example, $C_R \cup C_C$ has no solution since $\{CR_1, CR_2\} \cup C_C$ and $\{CR_1, CR_3\} \cup C_C$ are inconsistent and therefore both $\{CR_1, CR_2\}$ and $\{CR_1, CR_3\}$ induce a conflict[21] with the given compatibility constraints. Conforming with the hitting set algorithm,[20] we have to resolve each of the given conflicts. A minimal repair for C_R (resulting in C'_R) is to change the requirement related to the willingness to take risks, i.e. $wr_c = medium$ which makes $C'_R \cup C_C$ consistent ($C'_R = \{wr_c = medium, id_c = shortterm, kl_c = beginner\}$).

Explanation of Solutions. For each product part of a solution calculated by a recommender application, a set of immediate explanations[22] is calculated, i.e. a set of explanations which are derived from those filter constraints which are responsible for the selection of a product (see, e.g. the filter constraint of Fig. 10.3). An explanation related to our example filter constraint CF_7: $kl_c = beginner \Rightarrow ri_p <> high$ could be *this product assures adequate return rates with a lower level of related risks*. Note that explanations are directly assigned to filter constraints.

User Modeling. Due to the heterogeneity of users, the Koba4MS environment includes mechanisms allowing the adaptation of the dialog style to

Recommender Knowledge Base (relevant parts)		
Compatibility Constraints(C_C):		
CC_1: $\neg(wr_c = \text{high} \wedge id_c = \text{shortterm})$		
CC_2: $\neg(kl_c = \text{beginner} \wedge wr_c = \text{high})$		
CC_3: $\neg(sl_c = \text{bonds} \wedge id_c = \text{shortterm})$		
$\Updownarrow$ inconsistent ($\{CR_1, CR_2\}, \{CR_1, CR_3\}$)		$\Updownarrow$ consistent
Customer		**Customer**
Requirements (C_R)	$\Rightarrow$	**Requirements** (C'_R)
CR_1: $wr_c = \text{high}$	*repair action* $\{CR'_1\}$	CR'_1: $wr_c = \text{medium}$
CR_2: $id_c = \text{shortterm}$		CR_2: $id_c = \text{shortterm}$
CR_3: $kl_c = \text{beginner}$		CR_3: $kl_c = \text{beginner}$

Fig. 10.6. Example: repair of customer requirements.

the users skills and needs.[23] The user interface relies on the management of a user model that describes capabilities and preferences of individual customers. Some of these properties are directly provided by the user (e.g. age, nationality, personal goals, or self-estimates such as knowledge about financial services), other properties are derived using personalization rules and scoring mechanisms[1,23] which relate user answers to abstract dimensions such as *preparedness to take risks* or *interest in high profits* (dimensions describing the users interests).

10.3. Evaluation

10.3.1. *Example application*

Koba4MS technologies have been applied in a number of commercial projects (see, e.g. Refs. 24 and 1). Figure 10.7 depicts example screenshots of an investment advisor implemented for the Hypo Alpe-Adria-Bank in Austria (www.hypo-alpe-adria.at). First, a set of questions is posed to a customer, i.e. preferences are elicited (a). The corresponding answers provided by the customer (customer requirements) serve as input for the calculation of a solution. In the case that no solution can be found by the recommender application, the calculation of repair alternatives is activated (b). After having selected one of the alternatives, the customer can continue the recommendation session. Finally, a set of alternative investment proposals is determined and presented to the customer (in our case, two portfolios have been identified) (c). For each alternative, a corresponding set of explanations is calculated, as to why a certain product suits the wishes and needs of a customer (d). Finally, product comparisons provide basic mechanisms to compare different products which are part of a recommendation result (e) where differences between the selected (reference) product and other products are clearly indicated (the definition of comparison rules is shown in the simple example of Fig. 10.4).

A number of additional applications have been implemented on the basis of Koba4MS, e.g. financial service recommenders for the Wuestenrot and the Fundamenta building and loan association (www.wuestenrot.at, www.fundamenta.hu), recommenders for www.quelle.at, one of the leading online selling environments in Austria, the digital camera advisor for www.geizhals.at, the largest price comparison platform in Austria, and the recommender application which supports students at the Klagenfurt University (www.uni-klu.ac.at) in the identification of additional financial support opportunities (e.g. grants). Experiences from two selected projects will be discussed in the following subsection.

10.3.2. Experiences from projects

Financial services advisor. In the case of the *Wuestenrot building and loan association,* financial service advisors have been developed with the goal to support sales representatives in the dialog with the customer. The recommenders have been integrated with an existing Customer Relationship Management (CRM) environment and are available for 1.400 sales representatives.[1] The motivation for the development of financial service recommender applications was twofold. First, *time efforts* related to the prepa-

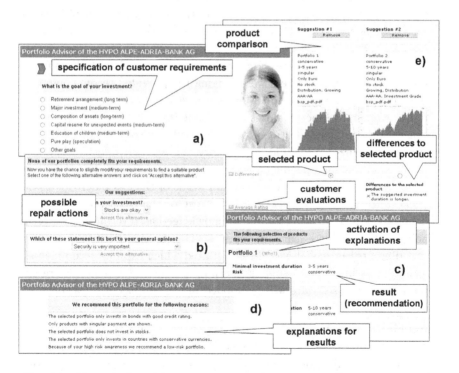

Fig. 10.7. Example financial services recommender application (investment advisor).

ration, conduction and completion of sales dialogues should be reduced. Second, an *automated documentation of advisory sessions* should be supported in order to take into account regulations of the European Union[25] related to the documentation of financial service advisory dialogs. With the goal to get a picture of how users apply the new technology and which impacts this technology has on sales processes, we have interviewed sales representatives of the Wuestenrot building and loan association ($n = 52$). Summarizing, the major results of the evaluation of the questionnaire were the following:[1,24]

- *Time savings*: On an average, interviewees specified the reduction of time efforts related to financial services advisory with 11.73 minutes per advisory session (SD (standard deviation) = 7.47), where the reductions are explained by the automated generation of advisory protocols and available summaries of previous dialogues at the beginning of a new advisory session. This corresponds to about 30% reduction of time efforts in the start phase and the final phase of an advisory session. Assuming that an average sales representative conducts about 170 advisory sessions per year, this results in time savings of about 33 hours per year.
- *Usefulness of recommender functionalities*: In the case of Wuestenrot the majority of interviewees were judging the provided recommendation functionalities as very useful. Most of the sales representatives reported to use Koba4MS functionalities throughout the sales dialogue or for the generation of documentations for completed advisory dialogues. Each such documentation includes a summary of customer preferences, a listing of recommended products, and an argumentation as to why the recommended products fit the wishes and needs of the customer.
- *Importance for new representatives*: Most of the sales representatives definitely agreed on the potential of knowledge-based recommender technologies to provide e-learning support. Due to this feedback, a corresponding project has already been initiated which exploits recommender technologies in order to support learning processes for new sales representatives. The software will be applied in the context of sales courses.

Financial support advisor. Apart from user studies in the financial services domain[1,24] we have evaluated the impacts of a financial support recommender application developed for students at the Klagenfurt University in Austria. On an average, about 150 students per month use the services of the financial support advisor for the identification of additional financial support opportunities. Our evaluation of the advisor consisted of $n = 1.271$ online users of www.uni-klu.ac.at, the homepage of the Klagenfurt University. An announced lottery ensured that participants identified themselves with their genuine names and addresses and no duplicate questionnaires were counted. The sample consisted of arbitrary online users of www.uni-klu.ac.at, who did not necessarily know the recommender application (36% if the participants already knew and 12% actively applied the advisor). The major results of this study were the following:

- *Increase of domain knowledge*: Of those interviewees who actively applied the recommender application, 69.8% reported a significant increase of domain knowledge in the area of financial support for students as a direct consequence of having interacted with the advisor.
- *Additional applications*: Of those interviewees who actively applied the recommender application, 19.4% applied for additional financial support as a direct consequence of having interacted with the advisor.
- *Time savings*: On an average, interviewees specified the overall time savings caused by the application of the advisor with 61.93 minutes per advisory session $(SD = 27.8)$. Consequently, students invest less time to get informed about additional financial support opportunities. Furthermore, members of the students council invest less time in routine advisory tasks.

10.3.3. Empirical findings regarding user acceptance

In this section we focus on the presentation of the results of a user study $(N = 116)$ which investigated explicit and implicit feedback of online users to various interaction mechanisms supported by knowledge-based recommender applications. The findings of the study show interesting patterns

Table 10.1. Different versions of *Internet Provider* advisors.

Advisor Versions
(a) switched expertise, positively formulated explanations, with product comparisons.
(b) switched expertise, without explanations, without product comparisons.
(c) positively formulated explanations, without product comparisons.
(d) negatively formulated explanations, without product comparisons.
(e) positively formulated explanations, with product comparisons.
(f) without explanations, with product comparisons.
(g) pure list of products (without any recommendation functionalities).
(h) positively formulated explanations, with product comparisons (automatically activated).

Table 10.2. Variables assessed in the study.

(a) Questions Posed Before Advisor Has Been Started

(1) previous usage (for buying purposes, as an information source)
(2) satisfaction with recommendation processes (advisory support) up to now
(3) trust in recommended products up to now (products suit personal needs)
(4) knowledge in the Internet Provider domain
(5) interest in the domain of Internet Providers

(b) Questions Posed After Completion of Advisory Session

(1) knowledge in the Internet Provider domain
(2) interest in the domain of Internet Providers
(3) satisfaction with the recommendation process (advisory support)
(4) satisfaction with the recommended products
(5) trust in the recommended products (products suit personal needs)
(6) correspondence between recommendations and expectations
(7) importance of explanations
(8) competence of recommender application
(9) helpfulness of repair actions
(10) willingness to buy a product

(c) Data Derived From Interaction Log

(1) session duration
(2) number of visited web pages
(3) number of inspected explanations
(4) number of activated product comparisons
(5) number of clicks on product details
(6) number of activations of repair actions

of consumer buying behavior when interacting with knowledge-based recommender applications. In particular, there exist specific relationships between the type of supported interaction mechanisms and the attitude of the user w.r.t. the recommender application. In the study we analyzed the degree to which concepts such as explanations, repair actions and product comparisons influence the attitudes of online users towards knowledge-based recommender technologies. In the scenario of the study the participants had to decide which online provider they would select for their home internet connection. To promote this decision, eight different versions of an *Internet Provider* recommender application have been implemented. The participants of the study had to use such a recommender application to identify the provider which best suits their needs and to place a fictitious order. Each participant was randomly assigned to one version of the implemented recommender applications (an overview of the provided versions of recommender applications is given in Table 10.1). Before and after interacting with the advisor, participants had to fill out an online questionnaire [see Tables 2(a) and 2(b)]. Participation was voluntary and a small remuneration was offered. We were interested in the frequency, participants used a recommender application either to order products or as an additional information source [Table 10.2(a)]. Self-rated knowledge and interest in the domain of internet connection providers (Table 10.2(a)-4,5) was assessed on a ten-point scale before interacting with the recommender application. After solving the task of virtually buying a connection from an Internet Provider, the participants had to answer follow-up questions which were also assessed on a 10-point scale [Table 10.2(b)] except Table 10.2(b) where a probability estimate had to be provided. Additional variables have been extracted from interaction logs [Table 10.2(c)]. The inclusion of the variables depicted in Table 10.2 is based on a set of hypotheses which are outlined in the following together with the corresponding exploratory results. The participants of the user study were randomly assigned to one of the Internet Provider advisors shown in Table 10.1. If a participant was confronted with the advisor version (a) or (b) and answered the question related to his/her expertise with *expert* than he/she was forwarded to a path in the recommender process which was designed for the advisory of beginners (and vice-versa) — we denote this as *switched expertise*. This manipulation was used to test the hypothesis that a dialog design fitting to the knowledge level of the participants leads to a higher satisfaction with the recommender application. Note that *positive explanations* provide a justification as to why a product fits to a certain customer, whereas *negative explanations* provide a justification for the relaxation of certain filter constraints. *Product comparisons* were supported in two different ways: first, comparisons had to be explicitly activated by participants, second, the result page was automatically substituted by the product comparison page. Finally, a *pure product list*, i.e. product selection without any advisory support, was implemented by automatically navigating to the result page and displaying all available products.

We tested 116 participants with a mean age of $\bar{x} = 28.7$ SD (standard deviation) $= 9.78$ (33.6% female). 42.2% were recruited from the Klagenfurt University and 57.8% were nonstudents. Explanations were used by 29.2% of the participants, repair actions have been triggered in 6.9% of the cases. Finally, a product comparison was used by 32.8% of the participants.[b] To assess the significance of correlations and differences, nonparametric tests

---

[b]Note that the relative frequencies refer to participants who had the possibility to use the corresponding feature (explanations, repairs, product comparisons).

were used.[26] Because the assessed variables were either ordinal-scaled or violated the assumptions of normal distribution or homogeneity of variance (visited pages, session duration), the Mann-Whitney U-Test was used to compare two groups and the Kruskal-Wallis-H Test to assess differences between more than two groups. In the following, only significant results are reported, with α set to 0.05 for all subsequent tests. The corresponding z-values are provided to show the size of the effects.

There were clear differences between the eight versions of recommender applications. The most positive ratings related to trust in the recommended products (Table 10.2(b)-5) and satisfaction with the recommendation process (Table 10.2(b)-3) were provided by participants interacting with the versions (e) and (h), i.e. advisor versions with positively formulated explanations and a product comparison functionality. Let us now consider the relationship between the features in the different advisor versions and the participants impressions in more detail.

Recommender application versus pure product list. We have found recommender applications to be more advantageous with respect to most of the assessed variables [see Table 10.2(b)]. Participants using a recommender application were significantly more satisfied with the recommendation process ($z = -3.872; p < 0.001$) (Table 10.2(b)-3) and had a significant increase in satisfaction due to the interaction with the Internet Provider advisor ($z = -2.938; p < 0.01$) (Tables 10.2(a)-2, 10.2(b)-3). Participants' trust in that the application recommended the optimal solution was higher for those interacting with the recommender application compared to those confronted with a pure product list ($z = -3.325; p = 0.001$) (Tables 10.2(b)-5). Furthermore, participants stated that the final recommendation better fitted to their expectations than when they were confronted with a simple product list ($z = -3.872; p = 0.001$) (Tables 10.2(b)-6). Most interestingly, the increase of subjective product domain knowledge due to the interaction was higher when participants interacted with a recommender application ($z = -2.069; p = 0.04$) (Tables 10.2(a)-4, 10.2(b)-1). The estimated (subjective) probability to buy a product in a purchase situation was higher for those interacting with a recommender application than for those interacting with a pure product list ($z = -2.1; p < 0.01$). Actually, this mean probability was only $p = 0.19$ for participants confronted with a product list, suggesting that these participants estimated a real purchase of the selected product as rather unlikely.

Effects of providing explanations. The perceived correspondence between recommended products and expectations (Table 10.2(b)-6) as well as the perceived competence of the recommender application (Table 10.2(b)-8) were rated higher by participants provided with the *possibility to use explanations* ($z = -3.228; p < 0.01$ and $z = -1.966; p < 0.05$). Most importantly, these participants trust in recommended products clearly increased due to the interaction process ($z = -2.816; p < 0.01$) (comparing pre- to post-test, Tables 10.2(a)-3, 10.2(b)-5). There is a tendency that providing explanations leads to more satisfaction with the recommendation process ($z = -1.544; p = 0.06$) (Table 10.2(b)-3). However, as hypothesized before the study, the increase in the rated knowledge from pre- to post-test did not differ significantly between both groups (Tables 10.2(a)-4, 10.2(b)-1). Participants who had *actively* (!) inspected explanations express a higher correspondence between expected and recommended products ($z = -2.176; p = 0.01$) (Table 10.2(b)-6) and an increased interest in the product domain when comparing pre- to post-test ($z = -1.769; p < 0.05$) (Tables 10.2(a)-5, 10.2(b)-2). Participants who inspected explanations and had experience with applying recommender applications, showed a tendency to rate the importance of explanations higher (Table 10.2(b)-7). They showed more trust in the recommended products (Table 10.2(b)-5) and stated a higher interest in the product domain (Table 10.2(b)-2). This suggests that a certain degree of familiarity with recommender applications is necessary in order to optimally exploit explanations.

Exploring variables that may potentially influence the actual use of explanations, it was found that experience correlated with the degree of explanation use. Participants already having experience with recommender applications were more likely to use an explanation ($r = 0.23; p < 0.05$) (Table 10.2(c)-3). Interpreting interaction processes with advisors as processes of preference construction, as described by Ref. 27 we assume that explanations influence preferences by adjusting the expectations of customers. This influence may be simply due to the fact that an explanation contains product features to which customers are primed. As argued in Ref. 27, priming of features causes customers to focus attention to those features and thus possibly to compare the recommended products with their expectations mainly along the primed features. This provides an explanation as to why the perceived correspondence between recommended and expected products and trust is higher when providing explanations.

Effects of product comparisons. Participants using recommender applications supporting product comparisons were more satisfied with the recommendation process ($z = -2.186; p = 0.03$) (Table 10.2(b)-3) and the recommended products ($z = -1.991; p < 0.05$) (Table 10.2(b)-4) than participants using advisors without product comparison support. Furthermore, participants using advisors with product comparisons showed a significant higher trust in the recommended products ($z = -2.308; p = 0.02$) (Table 10.2(b)-5). Product comparison functionality leads to a higher perceived competence of the recommender application ($z = -1.954; p < 0.05$) (Table 10.2(b)-8). Interacting with advisors supporting product comparisons leads to a clear increase in trust ($z = 3.016; p < 0.01$) (Tables 10.2(a)-3, 10.2(b)-5) and interest in the product domain (Internet Providers) ($z = 1.885; p < 0.05$) (Tables 10.2(a)-5, 10.2(b)-2). Interestingly, these positive effects seem to be due to the offer of comparisons and not to their usage since only 32.8% of the participants actually used them.

Those participants who actually used product comparisons, were more satisfied with the recommendation process ($z = 2.175; p = 0.03$) (Table 10.2(b)-3). Positive effects due to the possibility of using a product comparison were even accentuated for those participants who already had experiences with recommender applications. They were more satisfied with the suggested products ($z = 2.233; p = 0.03$) (Table 10.2(b)-4) and established more trust ($z = -3.658; p < 0.001$) (Table 10.2(b)-5). Furthermore, product comparisons combined with existing experiences leads to a higher perceived competence of the advisor ($z = 1.940; p < 0.05$) (Table 10.2(b)-8).

The multitude of positive influences that product comparisons offer (especially the increase in satisfaction) can be explained by the lower mental workload when products and product features are visually clearly presented to enable an evaluation of the recommended product set. Interestingly, taken together with the results on the explanation feature, some suggestions for the optimal design of product comparisons can be made. First, as already suggested by Ref. 28, it is useful for customers to visually highlight feature (settings) in the result that vary between the products (e.g. different color or font size). Also, assuming that a customers product evaluation will be rather based on features that she/he was primed to in the course of the interaction process through questions or an explanation feature, it should aid her/his purchase decision when primed features are highlighted as well. These implications will be tested in a follow-up study.

Effects of repair actions. [c] If we compare the participants who

---

[c] In the present study only 6.9% of the participants triggered repair actions. For this reason we combined the data with a sample from a pilot study.

triggered repair actions (due to their inconsistent specifications) to those who did not trigger repair actions, we find that the first group stated to have less knowledge in the product domain ($z = -1.801; p < 0.05$) (Table 10.2(a)-4) and that they rarely used recommender applications before ($z = -1.645; p < 0.05$) (Table 10.2(a)-1). This is plausible since participants with higher product domain knowledge and more experience with recommender applications will have more realistic expectations regarding product features and costs and they will provide information to an advisor that will most likely generate a set of recommended products, which makes a repair action dispensable. Thus, participants who used repair actions indicated an increase in product domain knowledge ($z = -1.730; p < 0.05$) (Tables 10.2(a)-4, 2(b)-1) and rated repair actions as more useful ($z = -2.978; p < 0.01$) (Table 10.2(b)-9).

Effects of *switched expertise*: Participants who received switched versions showed less satisfaction with the recommendation processes ($z = -1.790; p < 0.05$) (Table 10.2(b)-3) and provided a lower rating for the competence of the advisor ($z = -2.997; p < 0.01$) (Table 10.2(b)-8). They regarded the helpfulness of repair actions as lower ($z = -2.379; p < 0.01$) (Table 10.2(b)-9) compared to participants not confronted with the switched expertise scenario. This may be interpreted as an indicator of lower interest in recommender applications that fail to put questions that appropriately incorporate the expertise or knowledge level of the customer.

Willingness to buy a product: We examined which of the assessed variables show a significant correlation with the willingness to buy a product. The highest correlation has been detected between the willingness to buy (Table 10.2(b)-10) and trust in the recommended products ($r = 0.60; p < 0.01$) (Table 10.2(b)-5).[d] Furthermore, the higher the fit between the suggested products and the expectations of the participants (Table 10.2(b)-6), the higher was the willingness to buy the recommended product ($r = 0.54, p < 0.01$). Another interesting relationship exists between the perceived competence of the recommender application (Table 10.2(b)-8) and the willingness to buy ($r = 0.49, p < 0.01$) (Table 10.2(b)-10).

---

[d] For the computation of correlation measures, the Spearman correlation r for ordinal scale variables was used.

10.4. Related Work

Recommender Technologies. In contrast to collaborative filtering[9,11,12] and content-based filtering[13] approaches, knowledge-based recommendation[1,7,14,15] exploits deep knowledge about the product domain in order to determine solutions suiting the customers wishes and needs. Using such an approach, the relationship between customer requirements and products is explicitly modeled in an underlying knowledge base. Thus ramp-up problems[7] are avoided since recommendations are directly derived from user preferences identified within the scope of the requirements elicitation phase. The main reason for the choice of a knowledge-based recommendation approach stems from the requirements of domains such as financial services where deep product knowledge is needed in order to retrieve and explain solutions.[14] embed product information and explanations into multimedia-enhanced product demonstrations where recommendation technologies are used to increase the accessibility of the provided product descriptions. Using such representations, basic recommendation technologies are additionally equipped with a level supporting the visualization of calculated results. Reference 15 focused on the integration of conversational natural language interfaces with the goal of reducing system-user interactions. A study in the restaurant domain[15] clearly indicates significant reductions in efforts related to the identification of products (in terms of a reduced number of interactions as well as reduced interaction times). Natural language interaction as well as visualization of results are currently not integrated in the Koba4MS environment but are within the scope of future work. Compared to other existing knowledge-based recommender approaches,[7,14,15] Koba4MS includes model-based diagnosis[19,20] concepts allowing the calculation of repair actions in the case that no solution can be found and provides a graphical development environment which makes the development of recommender applications feasible for domain experts.[1]

User Acceptance of Recommender Technologies.[29] Evaluates navigational needs of users when interacting with recommender applications. A study is presented which reports results from an experiment where participants had to interact with recommender applications providing two different types of products (digital cameras and jackets offered in a digital store). It has been shown that different types of products trigger different navigational needs. The major factors influencing the navigational behaviour is the product type, e.g. compared to digital camera shoppers, jacket shoppers spent significant less time investigating individual products. The study in

Ref. 29 focused on the analysis of different navigational patterns depending on the underlying product assortment. The results presented in this paper report experiences related to the application of basic recommender technologies in online buying situations. The investigation of differences related to different product domains is within the scope of future work. Reference 27 analyzed the impact of personalized decision guides to different aspects of online buying situations. An interesting result of the study was that consumers choices are mostly driven by primary attributes that had been included in the recommendation process which clearly indicated the influence of personalized decision guides on consumer preferences. Compared to the work presented in this paper, Ref. 27 did not investigate effects related to the application of knowledge-based recommender technologies such as explanations of calculated results or repair actions. Furthermore, no detailed analysis has been done on psychological aspects of online buying situations such as trust, subjective perceived increase of domain knowledge, or the probability to buy a product. Reference 30 analyzed different dimensions of the users perception of a recommender agents trustworthiness. The major dimensions of trust which are discussed in Ref. 30 are systems features such as explanation of recommendation results, trustworthiness of the agent in terms of, e.g. competence and finally trusting intentions such as intention to buy or intention to return to the recommender agent. Where the results are comparable, the study presented in Ref. 30 confirms the results of our study (explanations are positively correlated with a user's trust and well-organized recommendations are more effective than a simple list of suggestions).

10.5. Summary and Future Work

We have presented the Koba4MS environment for the development of knowledge-based recommender applications. Koba4MS is based on innovative AI technologies which provide an intuitive access to complex products for customers as well as for sales representatives. Innovative technologies are crucial for successfully deploying recommender applications in commercial environments. However, a deeper understanding of the effects of these technologies can make recommender applications even more successful. A step towards this direction has been shown in this paper by analyzing the effects of mechanisms such as explanations, repair actions or product comparisons on a user's overall acceptance of the recommender application. The major direction of future work is the *integration of psychological the-*

ories from the area of consumer buying behavior into design processes of knowledge-based recommender applications. A corresponding project has already been started.

References

1. A. Felfernig and A. Kiener. Knowledge-based Interactive Selling of Financial Services using FSAdvisor. In *17th Innovative Applications of Artificial Intelligence Conference (IAAI'05)*, pp. 1475–1482, Pittsburgh, Pennsylvania, (2005).
2. L. Ardissono, L. Console, and I. Torre, An adaptive system for the personalized access to news, *AI Communications.* **14**(3), 129–147, (2001).
3. P. Resnick, N. Iacovou, M. Suchak, P. Bergstrom, and J. Riedl. GroupLens: An Open Architecture for Collaborative Filtering of Netnews. In *ACM Conf. on Computer Supported Cooperative Work*, pp. 175–186, (1994).
4. M. Montaner, B. Lopez, and J. D. la Rose, A Taxonomy of Recommender Agents on the Internet, *Artificial Intelligence Review.* **19**, 285–330, (2003).
5. J. Schafer, J. Konstan, and J. Riedl, Electronic Commerce Recommender Applications, *Journal of Data Mining and Knowledge Discovery.* **5**(1/2), 115–152, (2000).
6. G. Adomavicius and A. Tuzhilin, Toward the Next Generation of Recommender Systems: A Survey of the State-of-the-Art and Possible Extensions, *IEEE Transactions on Knowledge and Data Engineering.* **17**(6), 734–749, (2005).
7. R. Burke, Knowledge-based Recommender Systems, *Encyclopedia of Library and Information Systems.* **69**(32), (2000).
8. R. Burke, Hybrid Recommender Systems: Survey and Experiments, *User Modeling and User-Adapted Interaction.* **12**(4), 331–370, (2002).
9. B. Sarwar, G. Karypis, J. A. Konstan, and J. Riedl. Item-based collaborative filtering recommendation algorithms. In *10th Intl. WWW Conference*, pp. 285–295, (2001).
10. L. Terveen and W. Hill, *Beyond recommender systems: Helping people help each other, HCI in the New Millennium.* (Addison Wesley, 2001).
11. J. L. Herlocker, J. A. Konstan, L. G. Terveen, and J. Riedl, Evaluating Collaborative Filtering Recommender Systems, *ACM Trans. on Inf. Systems.* **22**(1), 5–53, (2004).
12. B. Smyth, E. Balfe, O. Boydell, K. Bradley, P. Briggs, M. Coyle, and J. Freyne. A Live User Evaluation of Collaborative Web Search. In *19th International Joint Conference on AI*, pp. 1419–1424, Edinburgh, Scotland, (2005).
13. M. Pazzani and D. Billsus, Learning and Revising User Profiles: The Identification of Interesting Web Sites, *Machine Learning.* (27), 313–331, (1997).
14. B. Jiang, W. Wang, and I. Benbasat, Multimedia-Based Interactive Advising Technology for Online Consumer Decision Support, *Comm. of the ACM.* **48**(9), 93–98, (2005).

15. C. Thompson, M. Goeker, and P. Langley, A Personalized System for Conversational Recommendations, *Journal of Artificial Intelligence Research.* **21**, 393–428, (2004).
16. A. Felfernig and K. Shchekotykhin. Debugging User Interface Descriptions of Knowledge-based Recommender Applications. In eds. C. Paris and C. Sidner, *ACM International Conference on Intelligent User Interfaces (IUI'06)*, pp. 234–241, Sydney, Australia, (2006).
17. P. Godfrey, Minimization in Cooperative Response to Failing Database Queries, *International Journal of Cooperative Information Systems.* **6**(2), 95–149, (1997).
18. D. McSherry. Maximally Successful Relaxations of Unsuccessful Queries. In 15th *Conference on Artificial Intelligence and Cognitive Science*, pp. 127–136, (2004).
19. A. Felfernig, G. Friedrich, D. Jannach, and M. Stumptner, Consistency-based Diagnosis of Configuration Knowledge Bases, *AI Journal.* **2**(152), 213–234, (2004).
20. R. Reiter, A theory of diagnosis from first principles, *AI Journal.* **23**(1), 57–95, (1987).
21. U. Junker, QUICKXPLAIN: Preferred Explanations and Relaxations for Over-Constrained Problems, *19th National Conf. on AI (AAAI04).* pp. 167–172, (2004).
22. G. Friedrich. Elimination of Spurious Explanations. In eds. G. Mueller and K. Lin, 16th *European Conference on AI (ECAI 2004)*, pp. 813–817, Valencia, Spain, (2004).
23. L. Ardissono, A. Felfernig, G. Friedrich, D. Jannach, G. Petrone, R. Schfer, and M. Zanker, A Framework for the development of personalized, distributed web-based configuration systems, *AI Magazine.* **24**(3), 93–108, (2003).
24. A. Felfernig, K. Isak, and C. Russ. Knowledge-based Recommendation: Technologies and Experiences from Projects. In 17th *European Conference on Artificial Intelligence (ECAI06)*, p. to appear, Riva del Garda, Italy, (2006).
25. E. Union, *Richtline 2002/92/EG des Europischen Parlaments und des Rates vom 9. Dezember 2002 ueber Versicherungsvermittlung.* (Amtsblatt der EU, 2002).
26. M. Hollander and D. Wolfe, *Nonparametric Statistical Methods.* (Wiley, 1999).
27. K. Murray and G. Haeubl, *Processes of Preference Construction in Agent-Assisted Online Shopping, in: Online Consumer Psychology.* (Lawrence Erlbaum, 2005).
28. W. Edwards and B. Fasolo, Decision technology, *Annual Review of Psychology.* (52), 581–606, (2001).
29. S. Spiekermann. Product Context in EC Websites: How Consumer Uncertainty and Purchase Risk Drive Navigational Needs. In *ACM Conference on E-Commernce 2004 (EC04)*, pp. 200–207, (2004).

30. L. Chen and P. Pu. Trust Building in Recommender Agents. In *International Workshop on Web Personalization, Recommender Systems and Intelligent User Interfaces (WPRSIUI05)*, pp. 135–145, Reading, UK, (2005).

BIOGRAPHIES

Alexander Felfernig received a Ph.D. degree in computer science from Klagenfurt University, Austria, in 2001. He is a co-founder of ConfigWorks, a provider of knowledge-based recommender technologies.

Bartosz Gula received the M.Sc. degree in psychology from the University of Berlin, Germany, in 2003. Currently, he is working as scientific researcher at Klagenfurt University.

Erich Teppan received the M.Sc. degree in computer science from Klagenfurt University, Austria, in 2005. Currently, he is working as scientific researcher at Klagenfurt University.

Chapter 11

Using Restricted Random Walks for Library Recommendations and Knowledge Space Exploration

Markus Franke and Andreas Geyer-Schulz

Information Systems and Management
Universität Karlsruhe (TH), Kaiserstr. 12
76131 Karlsruhe, Germany
maf@em.uni-karlsruhe.de

Implicit recommender systems provide a valuable aid to customers browsing through library corpora. We present a method to realize such a recommender especially for, but not limited to, libraries. The method is cluster-based, scales well for large collections, and produces recommendations of good quality. The approach is based on using session histories of visitors of the library's online catalog in order to generate a hierarchy of nondisjunctive clusters. Depending on the user's needs, the clusters at different levels of the hierarchy can be employed as recommendations. Using the prototype of a user interface we show that, if, for instance, the user is willing to sacrifice some precision in order to gain a higher number of documents during a specific session, he or she can do so easily by adjusting the cluster level via a slider.

11.1. Motivation

Amazon.com's recommender system ("Customers who bought this item also bought...") is an excellent and well-known example for a recommender service. Such services provide an added value for customers and sellers involved, for example:

- The *customers* or *users* receive assistance in browsing a product database they do not yet know the structure of and often get valuable tips as to which items may complement the current selection — items that might have been missed using conventional search strategies.
- In a commercial setting, the *seller* can increase his sales by offering the service; in a noncommercial setting like in a university library, e.g. user

satisfaction and thus perceived service quality can be increased by recommending books that are available.

In this contribution we are going to focus on implicit recommender services. By implicit recommender services we mean recommender services that are based on the analysis of observed user behavior. They operate without the need for explicit user cooperation by analyzing the traces generated by the users' sessions with the web interface of the shop or library catalog. Thus, contrary to explicit recommender systems which are based on surveying user intentions, opinions, or valuations by questionnaires, they are less prone to incentive-related problems like free riding or manipulation as discussed by Geyer-Schulz et al.[1] or Nichols.[2]

In the next section, we give a brief overview over the literature both on recommender systems and on cluster algorithms. We then start Sec. 11.3 with a description of the data set from the university library at Karlsruhe that is transformed into a similarity graph, and present the restricted random walk cluster algorithm. The algorithm is applied to the weight matrix of the similarity graph, resulting in a cluster hierarchy that can then be used to generate recommendations as shown in Sec. 11.4. Section 11.5 gives examples of the recommendations that can be generated by our method as well as some evaluations. We briefly address the issue of updating the recommendations when the similarity graph changes in Sec. 11.6. Finally, in Sec. 11.7, a short summary and an outlook are presented.

11.2. Cluster Algorithms and Recommender Systems

In this section, we present a short overview of the literature on both clustering and recommender systems.

For a general overview on recommender systems, we refer the reader to the articles by Gaul et al.,[3] Resnick and Varian,[4] Schafer et al.,[5] or Adomavicius and Tuzhilin.[6] The recommender systems that are surveyed in this section have in common that they are implicit and work on user protocol data like purchase histories, bookmarks, or newsgroup postings. They do not rely on content analysis of any kind, in contrast to the methods from information retrieval, as discussed for instance by Semeraro,[7] Yang,[8] and others.

For implicit recommendations, amazon.com uses algorithms based on weighted purchase correlations as described by Linden et al.,[9] and for explicit recommendations, amazon.com employs the collaborative filtering approach first presented in GroupLens with its own proprietary system architecture.[10,11] Although these algorithms are excellent examples of successful implicit and explicit recommender systems, they are by no means the only ones. The basic economic insight exploited by implicit recommender systems is that observed choice behavior reveals preference as stated by Samuelson.[12,13] The common principle behind implicit recommender systems is the following: first, scan the web server's log for user sessions. A session is either defined by an explicit login-logout cycle or by using the association of a user to a certain IP address for a given time, an approach that is used in the current implementation in Karlsruhe.[14] Inside the session, all purchases or viewings of a product's page are marked as one occurrence for this product. Based on these sessions, cross-occurrences between pairs of products are established if the two products occurred together in at least one session. These cross-occurrences are counted and summarized in a cross-occurrence matrix. In the second step, this matrix is evaluated in order to generate the recommendations. It is this step that constitutes the differences between the methods. Amazon's service works by simply recommending those items that have had the highest number or the highest share of cross-occurrences with the product in question. These basic recommendations can be modified by the product manager in accordance with special promotions or other motives.

More sophisticated systems like the one currently in use at the university's library in Karlsruhe are based on an adaption of repeat-buying theory and make use of the underlying theoretical distribution of independent purchase processes in a Poisson framework.[14,15] The stochastic framework makes them more robust and allows the detection of nonrandom cross-occurrences or the filtering of random cross-occurrences, dually. By a random cross-occurrence we mean that a basket contains products generated by independent stochastic purchase processes. A nonrandom cross-occurrence on the contrary is due to dependent purchase processes. Using the stochastic framework, nonrandom cross-occurrences can be identified by testing for outliers.

For a survey of clustering, we refer the reader to Refs. 16–18. The application of cluster methods for recommender systems and collaborative filtering has already been proposed in the past. A general survey of nonstandard techniques in collaborative filtering was compiled by Griffith and O'Riordan.[19] Among the first publications on the use of clustering techniques in this area was the work by Borchers et al.,[20] stating that clustering items leads to an improved quality of the recommendations and further fosters scalability of recommender systems by reducing the data set's size. Ungar and Foster have presented a collaborative filtering system that is based on a statistical partitioning of items and users.[21] User and item clusters are connected in this model by link probabilities. They tested the EM method, repeated clustering with k-means, and Gibbs sampling which they found to work best for their data set. The contribution by Kohrs and Merialdo focuses especially on the sparsity of matrices that is typical for recommender or collaborative filtering systems that are either new or have to integrate new users.[22] Sarwar et al. scrutinized the question of scalability of recommender systems.[23] As a solution, they proposed a clustering of the users with the scalable neighborhood algorithm.

Typically, these approaches utilize clustering algorithms as a preprocessing step in order to reduce the data set size and thus to decrease the execution time of the core recommender algorithms. By contrast, the method presented here relies completely on clustering to generate recommendations.

As Viegener has shown, there are clearly meaningful data to be found in library data — a good reason to further investigate the subject.[24] On the other hand, the single linkage clustering algorithm employed for his study has two drawbacks. First, the data set is large and clustering it with the selected algorithm required the use of the supercomputer at Karlsruhe's computing center. Second, single linkage clustering is prone to bridging which results in an insufficient cluster quality. The bridging effect in clustering occurs when small clusters are merged to one huge cluster and the structure of the small clusters becomes invisible. For library users bridging implies that in the worst case all books of the library are contained in one cluster or in a few very large clusters with several tens of thousands of documents. As a consequence we require a cluster algorithm that scales well with large data sets and produces clusters of high quality for this application.

We evaluated Restricted Random Walk (RRW) clustering as introduced by Schöll and Schöll-Paschinger.[25] The algorithm is sufficiently fast, and the quality of the resulting clusters fulfills the requirements of a recommender system.

11.3. Restricted Random Walk Clustering

The cluster method used for the generation of the recommendations has first been developed by Schöll and Schöll-Paschinger as a stochastic method on metric spaces.[25] We have applied it to the usage data from the university's library and could state that the quality of the resulting clusters is very high while the algorithm is sufficiently fast on standard PC hardware to be easily executed.[26,27] The method consists of two consecutive stages that we will describe in this section: the walk and the cluster construction stage. In the walk stage, random walks are executed on the document set in such a way that with each step the similarity between consecutively visited documents increases in a strictly monotonic way. The walk ends when no further increase in similarity is possible. At that point another walk is started. After a sufficient number of walks has been accumulated, clusters are constructed based on two principles: First, document pairs occurring at a late step of the walk have a higher similarity than those occurring at an early one where the stochastic influence is still high. Second, insights gained from longer walks are more important than those from shorter walks.

11.3.1. *Library usage data and similarity graphs*

Since RRW clustering depends on either an object set with distance information or a similarity graph, we will present here in brief the generation of such a graph from the library's web server log data. By browsing through the library's Online Public Access Catalog (OPAC), users generate entries containing the identifiers of documents whose detail pages they have visited during a session. These entries constitute our occurrences as defined in the introduction. When the detail pages of these documents are viewed together in the same session, this is a cross-occurrence between the documents. The session data are stored in session baskets that in turn are aggregated in order to obtain raw baskets for each document. A raw basket for a given document contains a list of all documents with which it has cross-occurrences, along with the cross-occurrence frequency, i.e. the respective number of cross-occurrences. An example of such a raw basket is given in Table 1 for object C in the example graph shown in Fig. 11.1.

Assuming that a high cross-occurrence frequency between documents implies a high complementarity, we now interpret these frequencies as similarities for the construction of the similarity graph $G = (V, E, \omega)$. V is the set of documents with a sufficient usage history: We omit all documents that were not viewed at all or were only viewed as the only document in their respective session, since these are outliers for which no cluster construction and thus no recommendation is possible. If two documents i and j have a positive cross-occurrence frequency, $E \subseteq V \times V$ contains an edge

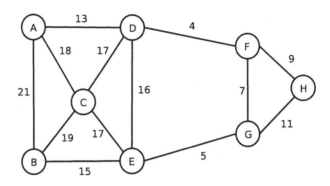

Fig. 11.1. An example graph.

between the two corresponding nodes. The weight on this edge, $\omega_{ij} = \omega_{ji}$ is set to the cross-occurrence frequency of i and j. In order to prevent the walk from looping over the same node repeatedly, ω_{ii} (the self-similarity) is set to zero. Finally, the neighborhood of a node i is defined as the set of nodes that share an edge with it.

11.3.2. *The restricted random walk stage*

A restricted random walk is a stochastic process leading to a series of visited states — in this case, nodes or edges of the similarity graph — that, contrary to a possibly infinite standard random walk, has a finite length. The formulation presented here differs in two aspects from the original one by Schöll and Schöll-Paschinger.[28] First, we are using similarities instead of distances which is uncritical since similarity measures can be easily converted to distance measures and vice versa.[29] The second point is more important insofar as the walks take place on the edges of the similarity graph instead of directly on the objects or nodes. This modification was

Table 11.1. Raw basket for object C in Fig. 11.1.

Object	Number of Cross-Occurrences
A	18
B	19
D	17
E	17

introduced in order to facilitate the analysis of the process with the means of Markov theory which we will introduce here in a short manner.

Consider a stochastic process visiting the states $\{i_0, i_1, \ldots\}$, where the i_m denoting the states of the process are realizations of stochastic variables I_m. Such a process is called a Markov chain if the probability of choosing j as successor state for i_{m-1} in the mth step is given by

$$P(I_m = j | I_0 = i_0, \ldots, I_{m-1} = i_{m-1}) = P(I_m = j | I_{m-1} = i_{m-1}), \quad (11.1)$$

i.e. the probability distribution over the possible successor states only depends on the current state.

Furthermore, the chain is called homogeneous if this probability distribution is independent of the step number. In this case, the transition probabilities can be written as $P = (p_{ij})$. An irreducible Markov chain is a chain in which each state can be reached from all other states via an arbitrary number of intermediary states.

The original formulation of the walk process leads to a finite and reducible second-order Markov chain; on the contrary, our formulation establishes an infinite and irreducible first-order Markov chain. With this, it becomes possible to study the behavior more closely for an infinite number of walks in order to obtain a clearer idea of the algorithm's properties.

To this end we define an additional state Ω of the process that is called the empty or transition state. This is the start state of the Markov chain, and the process returns to it every time a walk ends. From Ω, a first edge (i_0, i_1) of each walk is chosen from E with probability

$$P((i_0, i_1) | \Omega) = \frac{1}{|V| \deg(i_0)} \quad (11.2)$$

with $\deg(i_0)$ denoting the degree of node i_0.

This probability is derived from the original formulation of the restricted random walk,[28] where i_0 is chosen using a uniform distribution over all nodes $(1/|V|)$ and its successor i_1 is picked from a uniform distribution over all neighbors of i_0 $(1/\deg(i_0))$.

In each step $m + 1$ that is made during the walk a successor edge is chosen from a uniform distribution over the set

$$T_{i_{m-1} i_m} = \{(i_m, j) \in E | \omega_{i_m j} > \omega_{i_{m-1} i_m}\} \quad (11.3)$$

containing all edges incident to i_m with a higher similarity than the current one and which are therefore legal possible successors. At some point, a node will be reached via its highest-weighted incident edge. Per definition,

$T_{i_{m-1}i_m}$ is empty in this case, and the walk ends. The Markov chain once again enters Ω from where the next walk is started in an analogous fashion.

There is one aspect in which this formulation differs from the original one,[28] if one follows the idea of Schöll and Schöll-Paschinger for guaranteeing that every node is visited at least once. In order to achieve this, they have proposed to start a fixed number of walks from each node instead of picking the start node for each walk at random. For the practical implementation, this is not important since Eq. (11.2) can be replaced accordingly if one wishes to include each node as a starting node for a fixed number of walks. However, for the theoretical analysis one must bear in mind that the two are only equivalent asymptotically.

It is still an open question how the number of walks can be determined in order to obtain the desired cluster quality and reliability using as little computational effort as possible. Solutions for this problem are under development, both using random graph theory (cf. Ref. 30) and a theoretical analysis of the method's behavior for different numbers of walks including an infinite number of walks for stating the asymptotic behavior of the method that resembles a short-sighted (or local) single linkage clustering, but with less susceptibility to bridging.[31]

To get an idea of the procedure, consider the example graph in Fig. 11.1. Let us say that we obtain AD as first edge of the first walk. The probability of this happening is given by Eq. (11.2) as $1/(8*3) = 1/24$: We have eight nodes, from which we pick A with probability $1/8$. A in turn has three neighbors, so D is chosen with probability $1/3$. For the successor choice, we obtain $T_{AD} = \{DC, DE\}$, from which we pick, say, DE with probability $1/2$. Note that the edge DF is not a legal successor due to the similarity restriction that is used for the construction of $T_{i_{m-1}i_m}$ according to Eq. (11.3). The only possible successor for DE is given by $T_{DE} = \{EC\}$. Thus we pick, with probability one, EC. Again, we obtain a two-element successor set $T_{EC} = \{CA, CB\}$, from which CB is taken. For the next step, only BA can be selected since only this edge has a sufficiently high weight. After this, T_{BA} is empty which means that this walk ends here and the process enters the transition state Ω. Starting a walk from each of the nodes, we might get the walks $BEDCAB$, CAB, $DFHG$, $EGFHG$, $FDECBA$, $GEDCBA$, and HF.

Contrary to e.g. Single Linkage (SL) clustering, this method does not need to consider the whole similarity matrix at once. Only one row of the similarity matrix, containing local neighborhood information for the current node, is needed in each step, which greatly reduces space as well as time complexity and even allows to massively parallelize the walk stage without any efficiency losses since each walk is independent of the rest and even will be likely to access other areas of the similarity matrix than the other walks.

11.3.3. The cluster construction stage

As mentioned above, the first general idea used in the clustering stage is that the higher the position of a pair of nodes is in a walk, the higher is the probability that the two nodes are in the same cluster. This principle is used by both methods detailed here. We will first present the original one by Schöll and Schöll-Paschinger and discuss its properties before introducing the walk context approach.

However, before we start with this description, we should introduce the concept of hierarchical clustering. RRW clustering creates a hierarchy of clusters, such that at one end of the spectrum, only singletons, i.e. one-element clusters, exist; at the other end, all nodes are comprised in one big cluster. When moving from one of these extremes to the other, the clusters grow or shrink, depending on the direction. For disjunctive clusterings, this can be visualized by a dendrogram, setting the singletons at the bottom and the all-comprising cluster to the top. In order to obtain a cluster, it is necessary to fix a level l at which a horizontal "cut" is made through the dendrogram, creating a set of clusters at that level.

For our recommender application, we allow users to explore the cluster hierarchy for the document in question by enabling them to interactively change the level parameter according to their needs and to the quality and size of the list of recommended documents. Of course, it is equally feasible to offer the user only the m top-ranked recommendations by sorting the cluster members by the level at which they join or leave the cluster, and to give him/her the possibility of either requesting more recommendations or of further reducing the set if desired.

The component cluster construction method by Schöll and Schöll-Paschinger makes use of two series of graphs that are constructed on top of the accumulated walk data. This procedure is only based on the absolute position of a pair of nodes in the walk. As a first step, a series of graphs $G_k = (V, E_k)$ is created where V is the object set that is to be clustered and E_k contains an edge for each pair of nodes occurring in the kth step of any walk. Aggregating these, we obtain

$$H_l = \cup_{k=l}^{\infty} G_k \tag{11.4}$$

where the union of graphs consists of the union of their edge sets. A clustering at level l is then defined as the set of components or connected subgraphs of H_l. In other words, two nodes belong to the same cluster if and only if there is a path in H_l between them. As can easily be seen, the resulting clustering is disjunctive, i.e. each node belongs to exactly one cluster.

In our example containing the walks $ADECBA$, $BEDCAB$, CAB, $DFHG$, $EGFHG$, $FDECBA$, $GEDCBA$ and HF, we get the edge sets $V_1 = \{AC, AD, BE, DF, EG, FH\}$, $V_2 = \{AB, DE, FG, FH\}$, $V_3 = \{CD, CE, FH, GH\}$, $V_4 = \{AC, BC, GH\}$, $V_5 = \{AB\}$. Note that the edges are undirected, so we use AC also for CA, and that the members of the sets are sorted alphabetically for a better readability in this example. For step levels two and three, we obtain the clusters $\{A, B, C, D, E\}$, $\{F, G, H\}$ reflecting nicely the structure of the graph consisting of two clusters. At level four, the clusters separate into the singletons $\{D\}$, $\{E\}$, $\{F\}$, and the clusters $\{A, B, C\}$ and $\{G, H\}$ that form the denser parts of the clusters found at levels two and three. Finally at level five, only the cluster $\{A, B\}$ is found, the other nodes are contained in singleton clusters. Thus it can be stated that in this example, the clusters at different levels show very well the underlying hierarchical structure.

On the other hand, as will be discussed in the next section on recommendations, disjunctive clusters may not always be optimal for the task at hand. Furthermore, when applied to our library data set, component clusters proved to be too large, in some cases containing more than 100,000 documents even at the most restrictive level. This is most likely due to documents that were read in conjunction with other documents from different domains. One such document in a component cluster is sufficient to link two otherwise unrelated groups of documents, an effect well known as bridging.

In addition, the step number as level measure ignores the influence of the walk length on the reliability of the information derived from a walk. Consequently, the last step of a three-step walk and the third step of a ten-step walk are treated in the same way, even if in the latter case, the stochastic influence at the start of the walk is still strong. Furthermore, the step number depends on the course of the walk and cannot be fixed *a priori*.

In order to integrate both principles established at the beginning of Sec. 11.3, we tested some relative level measures. The first principle established at the beginning of this section, stating that a late position in a walk is more important than an early one, is, for instance, satisfied by the relative level

$$l = \frac{\text{step number}}{\text{total walk length}}. \qquad (11.5)$$

While this is definitely an improvement over the absolute step number, it ignores the second principle stipulating that the information from long walks is more valuable than that from short walks because the stochastic influence that is dominant at the walk's start diminishes with each step. However, the last steps both from a three-step walk and from a ten-step walk are treated with the same importance by the measure l. Additionally, one should bear in mind that the longer the walk, the finer is the resolution or the number of levels at which changes in the cluster structure take place. For instance, in a two-step walk, there are only three levels, 0, 0.5 and 1 available while in a ten-step walk, 11 levels can be used to differentiate cluster levels.

These considerations are taken into account by the measures

$$l^+ = \frac{\text{step number}}{\text{total walk length} + 1} \qquad (11.6)$$

and

$$l^- = \frac{\text{step number} - 1}{\text{total walk length}} \qquad (11.7)$$

that only asymptotically converge to one for infinitely long walks.

In addition to these alternative level measures, we developed the idea of walk context clusters. In order to construct a cluster for a given node (the central node), all walks are considered that contain that node at a level that is at least as high as the predefined cutoff. The cluster then consists of all nodes also occurring in these walks with a sufficiently high level. The idea is to reduce the influence of indirect connections that lead to cluster members only remotely similar to the central node, but that are nonetheless connected via one or several bridge elements. Using walk context clusters, the cluster size decreases considerably. Moreover the bridging effect is reduced because even if a bridge element is included in one of the walks, the number of nodes that are included due to this walk is limited to its members, and does not include all nodes accessible from the bridge element.

In our example, the clusters at $l^+ = 0.75$ are: $\{A, B\}$ for nodes A and B and $\{G, H\}$ for G and H with singleton clusters for the other nodes. At level $l^+ = 0.5$ the clusters are $\{A, B, C, D, E\}$ for A, B, and C, $\{A, B, C, D\}$ for

D, $\{A, B, C, E\}$ for E, and $\{F, G, H\}$ for F, G, and H. This is interesting, because due to the relatively "weak" connection between D and E as well as F and G, they are not included in each others' recommendation lists.

11.3.4. *Complexity*

One of the big strengths of RRW clustering is its low computational complexity compared to other cluster algorithms.

Let n denote the number of objects in V. According to Schöll and Schöll-Paschinger, the expected length of a single walk can be bounded by $O(\log n)$, leading to a total complexity of $O(n \log n)$ for all walks when a fixed number of walks is started from each node. Since the proof is based on the successive halving of the size of the successor set in each step, they propose to use two as base for the logarithm. If the number of neighbor nodes for a single node can be bounded by a constant, even a linear complexity is possible for the algorithm.[27]

For the second stage, the cluster construction, the complexity depends on the data structures used for storing the walks. For the complexity considerations, we will assume that the walk data are accessed via hash tables with a constant access time. On average, a node is visited by $O(\log n)$ walks: executing cn walks, where c is the number of walks started from each node, and with an expected walk length of $O(\log n)$, a total of $O(n \log n)$ entries is generated of which on average $O(\log n)$ walks visit each of the n nodes. With a bounded neighborhood size, this number is even constant. Thus, in order to retrieve the walks that contain the node in question, $O(\log n)$ steps are needed. Each of these walks comprises $O(\log n)$ nodes or $O(1)$ nodes, if the neighborhood size is bounded by a constant. So, the total complexity is $O(\log^2 n)$ for one cluster or $O(n \log^2 n)$ for all clusters, respectively $O(1)$ and $O(n)$ for the bounded neighborhood.

When applied to our library data set with ten walks per node the algorithm takes about two days on a standard PC with a dual Xeon cpu. The data set's graph contains 1.8 million nodes of which 800,000 have cross-occurrences with other nodes, and nearly 36 million weighted edges; the average degree of a node is about 39.

11.4. Giving Recommendations

With the clustering complete, the recommendation lists can easily be derived. Following our assumption about complementarities and their interpretation as similarity measure for the input of the cluster algorithm, we now reverse the argumentation: since the cluster algorithm finds groups with high intra-group similarity and since in our application this similarity in turn represents complementarities between documents, the clusters contain groups of books with high complementarity. This is exactly what is requested for a recommendation list.

As mentioned, the component cluster method produces disjunctive clusters, while the walk component clusters are nondisjunctive. The latter is a useful property when generating recommendations. Consider, for example, a document containing an introductory course on statistics for psychologists. Due to the nature of the construction of the similarity matrix, it will have a high similarity to books both from psychology and statistics. Consider now the clusters for the introductory course, and a book from one of the two areas, respectively. The first one should naturally contain documents from both areas, since the material forms a bridge between the subjects. On the other hand, the documents either from psychology or statistics may well be complemented by the introductory course, so it should be included into their clusters. But when creating the recommendation list for a book on psychology, the list of recommendations should not contain books on statistics in most cases. This requirement cannot be met

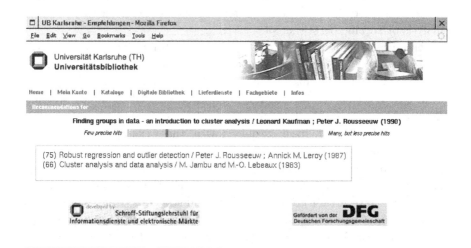

Fig. 11.2. Recommendations for Kaufman and Rousseeuw[29] with high precision, but low recall. The recommendations are also contained at the top of the list in Fig. 11.4.

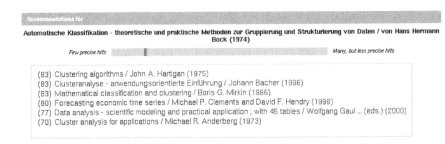

Fig. 11.3. Another example with high precision, this time for Bock's book.

by disjunctive clusters, whereas the nondisjunctive walk context clusters are fit to fulfill it.

However, the question remains how to set the value of l, the cutoff level for the hierarchical clustering. For the recommender application, the answer is surprisingly easy: in the prototypical interface, an initial default value on a medium level is used when the user opens the recommendation list. Afterwards, the user can adapt the cutoff level according to the needs of this specific search. For instance, the user might get something like Fig. 11.2 when calling up the recommendations for Kaufman and Rousseeuw's book.[29]

The default cutoff level for l^+ is set here to 0.6. As can be seen, the precision is good, both documents treat the subject classification/cluster analysis or specific parts thereof.

Another example for a cluster at level $l^+ = 0.7$ is shown in Fig. 11.3 for the book "Automatische Klassifikation" by Bock.[17] The fourth entry is not directly related to cluster analysis, but has strong ties to data analysis in that it treats the investigation and forecasting of economic systems described by measurable characteristics of these systems.

On the other hand, recall is quite low in these examples, so the user decides to accept also results with a possibly lower precision and sets the level to zero in order to obtain all possible recommendations. The resulting list is shown in Fig. 11.4. Going to the lower levels, there are more and more entries that have little to do with the original subject. Interesting to note is the group of marketing literature at the bottom of the list. Obviously, the inclusion of these can be explained by the use of cluster methods in marketing, for instance for the segmentation of consumers or markets.

As a result, recommendations generated by RRW clustering allow to delegate the decision on the tradeoff between precision and recall to where

> **Recommendations for**
>
> **Finding groups in data - an introduction to cluster analysis / Leonard Kaufman ; Peter J. Rousseeuw (1990)**
>
> *Few precise hits* ▬▬▬▬▬▬▬▬▬▬▬▬▬▬▬▬▬▬▬▬▬ *Many, but less precise hits*
>
> (75) Robust regression and outlier detection / Peter J. Rousseeuw ; Annick M. Leroy (1987)
> (66) Cluster analysis and data analysis / M. Jambu and M.-O. Lebeaux (1983)
> (50) Clusteranalyse - anwendungsorientierte Einführung / Johann Bacher (1996)
> (50) Empirical methods for artificial intelligence / Paul R. Cohen (1995)
> (50) Clustern mit Hintergrundwissen / Andreas Hotho (2004)
> (50) Bayeslösungen des Ausreißerproblems / Friedrich Gebhardt (1961)
> (50) Ausreisser bei ein- und mehrdimensionalen Wahrscheinlichkeitsverteilungen / Rudolf Mathar (1981)
> (33) Clustering algorithms / John A. Hartigan (1975)
> (33) Mathematical classification and clustering / Boris G. Mirkin (1996)
> (33) Data analysis - scientific modeling and practical application ; with 45 tables / Wolfgang Gaul ... (eds.) (2000)
> (33) Modern regression methods / Thomas P. Ryan (1997)
> (33) Untersuchung zur zeitlich-räumlichen Ähnlichkeit von phänologischen und klimatologischen Parametern in Westdeutschland u / von Xiaoqiu Chen (1994)
> (33) Cluster analysis / Brian Everitt (1974)
> (33) Social Science Research Council / Social Science Research Council ()
> (25) New approaches in classification and data analysis / E. Diday ... (eds.) (1994)
> (25) Clusteranalyse - Einführung in Methoden und Verfahren der automatischen Klassifikation ; mit zahlreichen Algorithmen, FO / Detlef Steinhausen ; Klaus Langer (1977)
> (25) Tests und Schätzungen in Ausreißermodellen / Ursula Gather (1984)
> (25) Concurrence probabilities for a locally slotted packet radio network by combinatorial methods / Rudolf Mathar (1990)
> (25) Classification and dissimilarity analysis / Bernard Van Cutsem (ed.) (1994)
> (25) Fallstudien Cluster-Analyse / Helmuth Späth (1977)
> (11) Mastering data mining - the art and science of customer relationship management / Michael J. A. Berry ; Gordon Linoff (2000)
> (11) Entwicklung von Kundenbeziehungen - theoretische und empirische Analysen unter dynamischen Aspekten / Dominik Georgi (2000)
> (11) Kundenwert - Grundlagen - innovative Konzepte - praktische Umsetzungen / Bernd Günter ... (Hrsg.) (2003)
> (11) Relationship Marketing - das Management von Kundenbeziehungen / von Manfred Bruhn (2001)
> (11) Customer-Lifetime-Value-Management - Kundenwert schaffen und erhöhen: Konzepte, Strategien, Praxisbeispiele / Markus Hofmann ... (Hrsg.) (2000)
> (11) Kundenwertmanagement - Konzept zur wertorientierten Analyse und Gestaltung von Kundenbeziehungen / Gunter Eberling. Mit einem Geleitw. von Günter Specht (2002)
> (11) Den Kundennutzen managen - so beschreiten sie den Weg zur Wertschöpfungskette / Harald Münzberg (1995)
> (8) Scheduling theory / Tanaev, Vjaceslav S. (1994)

Fig. 11.4. All recommendations for Kaufman and Rousseeuw: Recall is higher, but precision drops.

it best can be answered: to the user who is the only person who is aware of the specific search goal of the current session with the library's web interface.

As the lists in Figs. 11.3 and 11.4 show, an important advantage of behavior-based recommender systems is their independence of the document languages. In the context of scientific libraries this is very important because of the international nature of science and the resulting multilingual corpora.

11.5. Results

In lack of a human test group for an extensive direct evaluation of the resulting recommendation lists, we have performed two tests for the accuracy of the clusters in terms of library keywords. In addition, the results of a small user evaluation are provided in order to give a first impression of the quality of the recommendations. Finally, we discuss the applicability of other cluster algorithms to the library data set and give some evaluation results.

The first test uses the overlapping of keywords between the central node and the cluster members as a quality criterion.[32] We used the library's manual classification scheme that follows the SWD Sachgruppen introduced by the Deutsche Bibliothek to evaluate the clusters generated by restricted random walks for a sample of 40,000 documents.[33] A document is judged to be correctly classified into a cluster if it shares at least one category with the central node. The precision is then defined as

$$\text{precision} = \frac{\text{number of correctly classified documents}}{\text{number of documents in the cluster}}. \quad (11.8)$$

Recall could not be tested in a sensible way in this setting, since the manual classification as it is available at the library currently only covers about 55% of the documents. Consequently every computation of recall would be quite error-prone. Equally, this evaluation can only give a lower bound for the precision of the clusters, since document pairs that theoretically fit but of which at least one has not received a classification are counted as negatives.

As can be seen from Fig. 11.5, obtaining a clustering from a hierarchical clustering is a tradeoff between quality and quantity: at the left end of the scale, clusters of high precision can be generated. In our sample, a precision of 95% using l^+ was feasible, but at the cost of only obtaining clusters — and thus, recommendations — for 11 documents. If we wanted to e.g. generate recommendations for more than half of the documents, an average precision of 67% is feasible for 26,067 documents. With the introduction of the slider in the interface, this decision no longer has to be made by the administrators of the system, but the user can identify the optimal level in accordance with his individual and current needs.

A further point that should be noted here is the relative performance of the different level measures. The data shows that l is definitely dominated by the other two, while l^+ and l^- have no clear advantage over each other.

A similar picture is conveyed by the second evaluation that we conducted using RRW clustering for the generation of keywords in order to complement the manual index.[26] The general idea behind this algorithm is to construct clusters for documents that have no keywords in the manual classification and to assign those keywords to the central document that belong to a high share of documents from the cluster. This evaluation was conducted with reference librarians asked to judge keywords generated by the system for 212 documents at $l^+ = 0$. Nearly 4000 possible keywords were generated at this level of which 365 were judged to be correct for the respective document. This data set was used to investigate the dependency between precision, level and the number of keywords generated as plotted in Fig. 11.6. In order to compute whether a keyword is relevant for a given document we used a procedure that deviates from the original one:[28] instead of cutting off the clusters at a certain level and computing the share of documents in the cluster with a given keyword, we used the highest level at which a document is still a member of the cluster for the central document in question as an additional weight. The importance of a keyword k

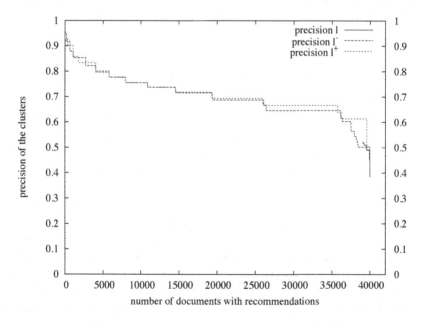

Fig. 11.5. Tradeoff between precision and number of documents in the recommendation list.

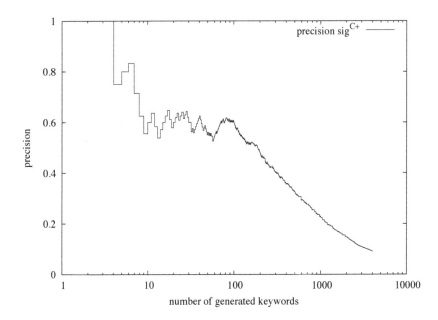

Fig. 11.6. The same tradeoff for recommended keywords for a set of 212 documents.

for a document i is then computed as

$$\mathrm{sig}^{C+}(k,i) = \frac{f_0(k,i)}{t_0(i)+1} \sum_{j \in C_0(i)} l^+_{\max}(i,j) \qquad (11.9)$$

where $j \in C_0(i)$ are the documents in i's cluster at level zero having keyword k, $t_0(i)$ is the cluster size at level zero, $f_0(k,i)$ is the number of documents at level zero that have the keyword k assigned, and $l^+_{\max}(i,j)$ is the maximum level at which document j belongs to document i's cluster. As can be seen, at the lowest level precision drops to about 10%. On the other hand, admitting only keywords with 100% precision, about 1% of the correct keywords could be assigned as can be seen from Fig. 11.6.

Finally, the authors evaluated the recommender lists for 30 documents from the domain of business administration and economy.[34] From these, the top five recommendations for each document were judged for complementarity to that document and accordingly sorted into the categories "good recommendation" or "not a recommendation". The results can be found in Table 11.2. An average precision of 78.66% could be achieved. In addition, more than 90% of the top five lists contained at least one useful recommendation.

For a general review of the quality of RRW clustering compared to other methods we refer the reader to the article by Schöll and Schöll-Paschinger.[25] They use two exemplary data sets to compare RRW, Single Linkage (SL) and Complete Linkage (CL) clustering. The first, containing elongated as well as compact clusters that are not well separated, shows that both RRW and CL are better suited than SL to discern clusters without clearly defined boundaries. The main difference between CL and RRW in this example is that RRW classifies objects as outliers that are included in the CL clusters. The second example consists of a ring-shaped cluster enclosing a compact spherical one. Both were correctly classified by the RRW method. SL and CL as well as group average methods all recognize four subgroups of the surrounding ring cluster, but merge these groups with the central compact cluster instead of with each other in order to obtain the original ring structure. For CL, this result is not surprising since CL is best suited for compact spherical clusters. By enlarging the distance between the ring and the compact cluster, at least SL was able to correctly identify the original clusters.

When conducting a direct evaluation of the suitability of other cluster methods for giving recommendations based on our library data set, two problems must be faced: the absence of a metric space and the sheer size of the data set. The library usage data is given as a sparse matrix of pairwise similarities between documents. Obviously, the triangle inequality does not necessarily hold in this case.[29] This means that e.g. centroid- or medoid-based methods could not be used. As a consequence we considered linkage-based algorithms, more specifically single and complete linkage, that are able to operate on similarities.[35]

Given the size of the data set with 800,000 objects and 36 million edges between them containing positive similarities, it would take much too long to compute all clusters and evaluate them in a fashion similar to the first evaluation cited above. But already when computing clusters for the two books used in Sec. 11.4, it becomes clear that both algorithms display

Table 11.2. Correct recommendations in the top five documents, sample size 30.

Correct Recommendations	5/5	4/5	3/5	2/5	1/5	0/5
Number of lists	17	5	3	1	2	2
Percentage	56.6	16.6	10.0	3.3	6.6	6.6

problems that are avoided by the RRW method with walk context clusters.

The raw basket for the book by Kaufman and Rousseeuw[29] contains 50 documents, but only four of them have a cross-occurrence frequency higher than one, the highest frequency being four. If we construct a cluster at this highest level, only taking into account edges with similarities of at least four, SL produces a cluster with nearly 250,000 documents. This is clearly too much for a recommendation list, especially when we take into account that the cluster only contains one of the original neighbors of the book.

On the other hand, the CL clusters suffer from a lack of differentiation in the list. While RRW clustering produced a total of seven different levels to choose from, CL naturally offers only four of them. Furthermore, when lowering the level from two to one, 46 out of the 50 total neighbors are added at once.

The situation is slightly better for the second example, the book by Bock,[17] where the highest similarity level is 19. The SL cluster at this level contains the books by Hartigan, Steinhausen, Mirkin, Bacher, and Gaul already known from Fig. 11.4. Once again, RRW offers 25 levels for the user to choose from, CL only nine, and similarly to the first example, 94 out of a total of 125 documents are added when including the most general level.

To sum it up, these examples show that if the distribution over the cross-occurrence frequencies is narrow and the maximum similarity is small, both SL and CL display a behavior that makes them less suited for the generation of recommendations than the RRW method. These effects are mitigated when contemplating a raw basket that contains a high maximum similarity and a broad distribution. It might be interesting to investigate some improvements of the linkage methods like a normalization of similarities, but this is outside the scope of this contribution. The most important argument against the use of linkage-based methods however is their computation time that is much higher than that of RRW clustering on the given library data set.

11.6. Updates of the Recommendation Lists

Another considerable advantage of the use of RRW clustering for recommendation purposes is the fact that it is relatively easy to incorporate changes in the raw baskets into the recommendation list.[36] These changes are due to the continuing use of the library's web interface, but also to inventory management measures like the acquisition of new books. By suitably truncating and reexecuting the walks affected by increased cross-occurrences, new books or the removal of old books from the library, it is possible to update the clusters with minimal effort. For k changes and n documents, the complexity of the updates is in the order of $O(k \log^2 n)$. Given the small volume of these changes — for a week, the baskets for about 10000 documents are updated — it is highly desirable to use an efficient update procedure that keeps the clusters and thus the recommendations current without the need for a complete reclustering.

11.7. Conclusion

We have presented an innovative approach for efficiently generating library recommendations based on user purchase histories with restricted random walks. The quality of the recommendations as well as the computational performance of the algorithm are promising.

For the next steps, we will make the interface public in order to gain user feedback on the recommendations and perform an evaluation based on the users' judgments. As for the theoretical side of the algorithm, measures for fine tuning the parameters will be scrutinized. The investigation of the influence of the number of walks that are started from each node as well as a better understanding of the asymptotic behavior of the walk process will help to improve the computational complexity while maintaining the cluster quality. As a further measure for increasing the quality of the clusters, we plan to separate the meaningful data from statistic influences as proposed by Geyer-Schulz et al.[14]

Acknowledgments

We gratefully acknowledge the funding of the project "RecKVK" by the Deutsche Forschungsgemeinschaft (DFG) as well as the funding of the project "SESAM" by the Bundesministerium für Bildung und Forschung (BMBF). We thank the editors, Gulden Uchyigit and Matthew Ma, and the publisher, World Scientific Publishing, for giving us the opportunity to publish a revised version of our paper.[37]

References

1. A. Geyer-Schulz, M. Hahsler, and M. Jahn, Educational and scientific recommender systems: Designing the information channels of the virtual university,

Int. J. Engin. Ed. **17**(2), 153 – 163, (2001).
2. D. M. Nichols. Implicit rating and filtering. In *Fifth DELOS Workshop: Filtering and Collaborative Filtering*, pp. 28–33. ERCIM, (1997).
3. W. Gaul, A. Geyer-Schulz, M. Hahsler, and L. Schmidt-Thieme, eMarketing mittels Recommendersystemen, *Marketing ZFP.* **24**, 47 – 55, (2002).
4. P. Resnick and H. R. Varian, Recommender Systems, *CACM.* **40**(3), 56 – 58, (1997).
5. J. B. Schafer, J. A. Konstan, and J. Riedl, E-commerce recommendation applications, *Data Min. Knowl. Discov.* **5**(1/2), 115–153, (2001).
6. G. Adomavicius and A. Tuzhilin, Toward the next generation of recommender systems, *IEEE Trans. Knowled. Data Engin.* **17**(6), 734 – 749, (2005).
7. G. Semeraro, S. Ferilli, N. Fanizzi, and F. Esposito. Document classification and interpretation through the inference of logic-based models. In ed. I. S. P. Constantopoulos, *Proc. 5th European Conf. ECDL 2001*, pp. 59–70, Heidelberg, (2001). Springer Verlag.
8. Y. Yang, An evaluation of statistical approaches to text categorization, *Inform. Retr.* **1**(1), 69–90, (1999).
9. G. D. Linden, J. A. Jacobi, and E. A. Benson. Collaborative recommendations using item-to-item similarity mappings. U.S. Patent 157198, assignee amazon.com, (2001).
10. J. Jacobi and E. Benson. System and methods for collaborative recommendations. U.S. Patent 6064980, assignee amazon.com, (1998).
11. P. Resnick, N. Iacovou, P. Bergstrom, and J. Riedl. GroupLens: An open architecture for collaborative filtering of netnews. In *Proc. Conf. Computer Supported Cooperative Work*, pp. 175 – 186, NY, (1994). ACM Press.
12. P. A. Samuelson, A note on the pure theory of consumer's behaviour, *Economica.* **5**(17), 61 – 71, (1938).
13. P. A. Samuelson, Consumption theory in terms of revealed preference, *Economica.* **15**(60), 243 – 253, (1948).
14. A. Geyer-Schulz, A. Neumann, and A. Thede, An architecture for behavior-based library recommender systems – integration and first experiences, *Inform. Technol. Lib.* **22**(4), 165 – 174, (2003).
15. A. S. Ehrenberg, *Repeat-Buying: Facts, Theory and Applications*. (Charles Griffin & Company Ltd., London, 1988).
16. J. C. Bezdek, *Pattern Recognition with Fuzzy Objective Function Algorithms*. (Plenum Press, NY, 1981).
17. H. Bock, *Automatische Klassifikation*. (Vandenhoeck & Ruprecht, Göttingen, 1974).
18. R. O. Duda, P. E. Hart, and D. G. Stork, *Pattern Classification*. (Wiley-Interscience, NY, 2001), 2 edition.
19. J. Griffith and C. O'Riordan. Non-traditional collaborative filtering techniques. Technical Report NUIG-IT-121002, National University of Ireland, Galway, (2002).
20. A. Borchers, D. Leppik, J. Konstan, and J. Riedl. Partitioning in recommender systems. Technical Report 98-023, University of Minnesota, Minneapolis, (1998).

21. L. H. Ungar and D. P. Foster. Clustering methods for collaborative filtering. In *Proc. Workshop on Recommendation Systems*, Menlo Park, (1998). AAAI Press.
22. A. Kohrs and B. Merialdo. Clustering for collaborative filtering applications. In *Computational Intelligence for Modelling, Control & Automation 1999*, pp. 199–204, Amsterdam, (1999). IOS Press.
23. B. M. Sarwar, G. Karypis, J. Konstan, and J. Riedl. Recommender systems for large-scale e-commerce: scalable neighborhood formation using clustering. In *Proc. Fifth Int. Conf. Computer and Information Technology*, Bangladesh, (2002).
24. J. Viegener, *Inkrementelle, domänenunabhängige Thesauruserstellung in dokumentbasierten Informationssystemen durch Kombination von Konstruktionsverfahren*. (infix, Sankt Augustin, 1997).
25. J. Schöll and E. Schöll-Paschinger, Classification by restricted random walks, *Patt. Recogn.* **36**(6), 1279–1290, (2003).
26. M. Franke and A. Geyer-Schulz. Automated Indexing with Restricted Random Walks on Large Document Sets. In eds. R. Heery and L. Lyon, *Research and Advanced Technology for Digital Libraries – 8th European Conf.*, ECDL 2004, pp. 232–243, Heidelberg, (2004). Springer.
27. M. Franke and A. Thede. Clustering of Large Document Sets with Restricted Random Walks on Usage Histories. In eds. C. Weihs and W. Gaul, *Classification – the Ubiquitous Challenge*, pp. 402–409, Heidelberg, (2005). Springer.
28. J. Schöll. *Clusteranalyse mit Zufallswegen*. PhD thesis, TU Wien, (2002).
29. L. Kaufman and P. J. Rousseeuw, *Finding Groups in Data*. (John Wiley, NY, 1990).
30. P. Erdös and A. Renyi, On random graphs I, *Publ. Math.* **6**, 290–297, (1957).
31. M. Franke and A. Geyer-Schulz. A method for analyzing the asymptotic behavior of the walk process in restricted random walk cluster algorithm. In eds. R. Decker and H.-J. Lenz, *Advances in Data Analysis*, pp. 51 – 58, Heidelberg, (2007). Springer.
32. M. Franke and A. Geyer-Schulz. Using restricted random walks for library recommendations. In ed. G. Uchyigit, *Web Personalization, Recommender Systems and Intelligent User Interfaces*, pp. 107 – 115, Setúbal, (2005). INSTICC Press.
33. M. Kunz and et al. SWD Sachgruppen. Technical report, Deutsche Bibliothek, Frankfurt/Leipzig, (2003).
34. M. Franke, A. Geyer-Schulz, and A. Neumann. Building recommendations from random walks on library opac usage data. In eds. S. Zani, A. Cerioli, M. Riani, and M. Vichi, *Data Analysis, Classification and the Forward Search*, pp. 235 – 246, Heidelberg, (2006). Springer.
35. G. N. Lance and W. T. Williams, A general theory of classificatory sorting strategies. I. Hierarchical systems, *Comp. J.* **9**(4), 373 – 380, (1967).
36. M. Franke. *An Update Algorithm for Restricted Random Walk Clusters*. PhD thesis, Universität Karlsruhe (TH), (2007).

37. M. Franke and A. Geyer-Schulz, Using restricted random walks for library recommendations and knowledge space exploration, *International Journal of Pattern Recognition and Artificial Intelligence.* **21**(2), 355 – 373, (2007).

BIOGRAPHIES

Markus Franke received his diploma in information engineering and management from the Research University Karlsruhe (TH) in 2003, from where he also received his Ph.D. in 2007. Currently, he is a research assistant at the Institute for Information Systems and Management in Karlsruhe.

Andreas Geyer-Schulz received his diploma in business administration from the Vienna University of Economics and Business Administration in 1982. He received his Ph.D. in 1985 and habilitated in 1995. After working as a professor in Augsburg and Vienna, he currently holds the Schroff chair for information services and electronic markets at the Institute for Information Systems and Management of the Universität Karlsruhe (TH).

Chapter 12

An Experimental Study of Feature Selection Methods for Text Classification

Gulden Uchyigit[*] and Keith Clark[†]

Department of Computing, Imperial College
180 Queen's Gate, South Kensington
London SW7 2AZ
[*]g.uchyigit@imperial.ac.uk
[†]k.clark@imperial.ac.uk

Text classification is the problem of classifying a set of documents into a pre-defined set of classes. A major problem with text classification problems is the high dimensionality of the feature space. Only a small subset of these words are feature words which can be used in determining a document's class, while the rest adds noise and can make the results unreliable and significantly increase computational time. A common approach in dealing with this problem is *feature selection* where the number of words in the feature space are significantly reduced.

In this paper we present the experiments of a comparative study of feature selection methods used for . Ten feature selection methods were evaluated in this study including the new feature selection method, called a *GU* metric. The other feature selection methods evaluated in this study are: Chi-Squared ($\chi 2$) statistic, NGL coefficient, GSS coefficient, Mutual Information, Information Gain, Odds Ration, Term Frequency, Fisher Criterion, BSS/WSS coefficient. The experimental evaluations show that the *GU* metric obtained the best F_1 and F_2 scores. The experiments were performed on the 20 Newsgroups data sets with the Naive Bayesian Probabilistic Classifier.

12.1. Introduction

The advent of the Internet, personal computer networks and interactive television networks has lead to an explosion of information available online from thousands of new sources. Much of this information is in the form of natural language texts. Hence, the ability to automatically classify this information into different categories is highly desirable. A major difficulty with textual data is the large number of words that can exist in the domain. Even for a medium sized document collection there can be tens or thousands of different words. This is too high for many classification algorithms. In order to improve scalability of the classification algorithms and reduce over-fitting, it is desirable to reduce the number of words used by the classification algorithm. Further, it is desirable to achieve such a goal automatically without sacrificing the classification accuracy. Such techniques are known as automatic feature selection methods. In general automatic feature selection methods on textual data include the removal of non-informative words and application of a feature scoring method to the remaining words. Only the top scoring words are then used as the significant words of the document set. The classification algorithms are then trained on this reduced feature set of significant words.

Our interest in feature selection stems form our research into TV recommender systems.[1] We have developed an agent-based system that comprises of personal assistant agents and collaboration agents. The personal assistant agents interact with their user to learn and continually modify the user's profile representing their viewing preferences. The profile essentially comprises, for each program category, e.g drama or comedy a weighted set of words. These word sets are culled from electronic program guides and reviews of programs. The profile changes as new feedback regarding viewed programs is given by the user. For the learning of the user profile the agents use the naive bayesian probabilistic classifier. These profiles are used to make viewing recommendations for new programs. These are recommendations based on past viewing habits and are the so called *Content-based recommendations*. But sometimes the user watches a program they liked but was not recommended by the system. When this happens it is useful to disseminate this positive feedback to other users that are similar to this user with respect to that type of program. Our system does this by dynamically grouping users into like minded interest groups for each program category.[2] These groups are constructed and maintained by collaboration agents who are sent the user profile when it is constructed and each time it is modified by each user agent.

The profiles comprise a relatively small set of feature words for each program category. Since these features sets have a pivotal role, it was essential for us to use the best feature selection method we could find. To this end we investigated and evaluated several proposed and previously used algorithms. This investigation lead us to develop a new algorithm which we present in this paper (the GU metric). We also present the results of our comparative evaluation. The new algorithm performs as well as or better than the other algorithms on a standard data set.

12.2. Text Classification

Text classification problems assign a document to a pre-defined set of categories. Each document can belong to one, more than one or no category at all. Using machine learning the objective is to learn to classify documents from examples. In general, this is a supervised learning task which requires a pre-classified set of training examples, in which each training example is assigned a unique label indicating the class it belongs to, among a finite set of possible set of classes. The goal of the classification algorithm is then to classify novel unlabeled examples.

The act of personalization, where the system automatically learns a user's interests and filters information on behalf of its user, may be seen as a classification problem where observation of user behavior can provide training examples to the classification algorithm which is used to form a model of the user's profile. This model is then to used to predict user's interests. In personalization at least two classes can be identified those pieces of information (i.e Web pages, news, e-mail messages, TV programs etc.) that are considered to be interesting and those that are considered as not being interesting. Most personalization tasks respond to the binary classification, for example a Web page can be classified as interesting or not interesting, an e-mail message can be classified as spam or not spam, and so forth. In order to build a user model these two classes act as positive and negative examples representing user interests. These examples are assimilated either implicitly or explicitly. The user can provide continuous feedback to guide the further learning of the system. User profiling can also be mapped to classification problems in which more than two classes are available, such as classifying e-mail messages into personal folders (e.g work, finances, personals etc.) or news into pre-defined news groups (e.g rec.autos, rec.sport.baseball, talk.politics.mideast etc.)

12.3. Text Representation

Text representation has been a long-standing endeavor of the information retrieval community. The classical models in information retrieval consider that each document is described as a set of keywords called index words. An index word is simply a word whose semantics help in remembering the document's main theme. Thus index words are used to index and summarize the document. In text classification vector representation of documents is a common representation technique.

Vector representation uses boolean or numerical vectors to represent the text documents. Each text document is viewed as a vector in n dimensional space, n being the number of different significant words in the document set. Such a representation is often referred to as *bag-of-words*, because word ordering and text structure are lost (see Fig. 12.1). The tuple of weights associated with each word, reflecting the significance of that word for a given document, give the document's position in the vector space. The weights are related to the number of occurrences of each word in the document.

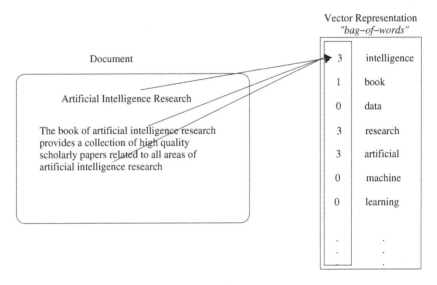

Fig. 12.1. Illustration of the *bag-of-words* document representation using word frequency.

12.4. Feature Selection

Often in textual domains the document vectors can grow to be very big containing several hundreds of features. In machine learning, theories put forward suggest that maximum performance is often not achieved using all available features, but using only a good subset of these. Applied to the text classification this means identifying a subset of words which helps to discriminate between classes. Having too few features make it impossible to formulate a good hypothesis. But having features which do not help discriminate between classes adds noise.

John et al.[3] described two main approaches to feature selection used in machine learning: the filter approach and a wrapper approach.

12.4.1. Filter Approach

In the filter approach (see Fig. 2), a feature is selected independently from the learning method that will use the selected features. This step happens before the induction step, where the irrelevant features are filtered before the induction process starts. There are methods in statistics that are used to reduce the number of features. Almuallim and Dietterich[4] developed several feature selection algorithms including a simple exhaustive search and algorithms that use different heuristics. They based their feature selection function on conflicts in class value occurring when two examples have the same values for all selected features.

12.4.2. Wrapper Approach

The wrapper approach (see Fig. 12.3) selects features using an evaluation function based on the same learning algorithm that will be used for learning on the domain represented with the selected features. The main disadvantage of the wrapper approach over the filter approach is the former's computational cost, which results from calling the induction algorithm for each feature considered.

12.5. Feature Selection for Textual Domains

In general, feature selection methods used for text learning and information retrieval is simpler than the feature selection methods used in machine learning . Learning on text defines a feature for each word that occurred in training documents. All features are independently evaluated (using a

Fig. 12.2. Feature filter model in which the features are filtered independent of the induction algorithm.

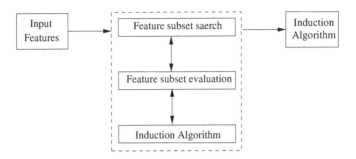

Fig. 12.3. The wrapper model.

feature scoring method), a score is assigned to each word and they are sorted according to the assigned score. Then, a predefined number of best scoring features is taken from the feature space. This paper is concerned with the different feature selection metrics which are used in feature selection for text classification.

Various feature selection metrics have been proposed in literature. Most of these methods try to capture the intuition that the most important words for classification are those that occur most frequently in either the set of relevant (positive), or the set of irrelevant (negative) examples of each category, and *not both*. For example consider the two sets of textual documents, one containing documents which are political news articles, and the other set containing documents which are sports articles. Suppose the word *politics* occurs in 80 out of the 100 political news articles whereas in the other set it does not occur at all. The word politics is therefore a good word for categorizing political news articles out of a set that contains these and sport articles. It is given a high score as a feature word for the politics category.

The next section describes the different feature scoring metrics which we have used in our experimental evaluations, including a new metric which is described at the end of the section.

12.6. Feature Scoring Metrics

The feature selection methods used in our experiments are: Chi-Squared Statistic $\chi 2$, Odds Ratio, Mutual Information, Information Gain, Document Frequency, NGL coefficient, GSS coefficient, Fisher criterion and BSS/WSS coefficient and finally the new GU metric.

Table 12.1. Contingency table representation of word w and categories c, and $\bar{c}$.

	Document belongs to category c c	Document does not belong to category c $\bar{c}$	
Document contains word w	c_w	$\bar{c}_w$	$c_w + \bar{c}_w$
Document doesn't contain word w	$c_{\bar{w}}$	$\bar{c}_{\bar{w}}$	$c_{\bar{w}} + \bar{c}_{\bar{w}}$
	$n_c = c_w + c_{\bar{w}}$	$n_{\bar{c}} = \bar{c}_w + \bar{c}_{\bar{w}}$	N

Throughout this section we will use the notation (from a two-way contingency table for word w and category c see Table 12.1) c_w, $\bar{c}_w$, $c_{\bar{w}}$, $\bar{c}_{\bar{w}}$, respectively, to denote: the number of times c and w co-occur; the number of times w occurs without c; the number of times c occurs without w; the number of times neither c nor w occurs. n_c is the total number of documents in c; $n_{\bar{c}}$ is the number of documents in $\bar{c}$ and N is the total number of documents in the collection (i.e. $n_c + n_{\bar{c}}$).

12.6.1. Chi-Squared statistic

The Chi-Squared (χ^2) statistic was originally used in statistical analysis to measure how the results of an observation differ (i.e. are independent) from the results expected according to an initial hypothesis (higher values indicate higher independence). In the context of text categorization χ^2 statistic is used to measure how independent a word (w) and a category (c) are.[5-7]

χ^2 has a value of zero if w and c are independent. A word which occurs frequently in *many* categories will have a low χ^2 value indicating high independence between w and c. In contrast, a word which appears frequently in *few* categories will have a high χ^2 value (i.e. high dependence). In our experiments we compute χ^2 using the equation below:

$$\chi^2_w = \frac{N \times (c_w \bar{c}_{\bar{w}} - c_{\bar{w}} \bar{c}_w)^2}{(c_w + \bar{c}_w) \times (\bar{c}_w + c_{\bar{w}}) \times (\bar{c}_w + \bar{c}_{\bar{w}}) \times (c_{\bar{w}} + \bar{c}_{\bar{w}})}$$

χ^2 is one of the metrics which has shown good performance in previous evaluation experiments.[5-7] The χ^2 statistic assigns high scores to words which are indicative of membership of category c but also those words which are indicative of non-membership of category c.

12.6.2. NGL coefficient

Ng et al.[8] proposed the Correlation Coefficient (CC), a variant of χ^2 metric, the square of CC is the χ^2 value, to be used in text classification. In our experiments we compute the NGL Coefficient using the equation below:

$$CC_w = \frac{\sqrt{N}(c_w \bar{c}_{\bar{w}} - c_{\bar{w}} \bar{c}_w)}{\sqrt{(c_w + c_{\bar{w}}) \times (\bar{c}_w + \bar{c}_{\bar{w}}) \times (c_w + \bar{c}_w) \times (c_{\bar{w}} + \bar{c}_{\bar{w}})}}.$$

A positive value of CC_w indicates w is a a possible feature word and correlates with c, while a negative value means w correlates with $\bar{c}$.

The NGL coefficient[8] is reported to have better performance than χ^2. Ng et al. report that NGL's better performance is due to the fact that it selects those words that correlate with c (i.e. are positive) and does not select those words which correlate with $\bar{c}$, unlike the χ^2 statistic.

12.6.3. GSS coefficient

Galavotti et al.[6] proposed a *simplified* χ^2 ($s\chi^2$) statistic. The simplification is made by: removing the $\sqrt{N}$ factor from the numerator (since it is equal for all pairs of c_w, and therefore becomes redundant); removing $\sqrt{(c_w + \bar{c}_w) \times (c_{\bar{w}} + \bar{c}_{\bar{w}})}$, since this has a low value for rare words and therefore resulting in rare words being given a high CC score (previous studies[25] have shown rare words to be least effective in text classification problems); removing the factor $\sqrt{(c_w + \bar{c}_w) \times (\bar{c}_w + c_{\bar{w}})}$ from the denominator since it serves to emphasize very rare categories (i.e categories with very few examples). Removing these three factors from CC yields the GSS coefficient. In our experiments we compute the GSS coefficient using the equation below:

$$s\chi^2 = (c_w \bar{c}_{\bar{w}} - c_{\bar{w}} \bar{c}_w).$$

Galavotti et al.[6] show that GSS coefficient outperforms both the χ^2 statistic and the NGL coefficient. Similar to NGL, the positive values correspond to features which correlate with c, while negative values correspond to features which correlate with $\bar{c}$, and so only considring positive words.

12.6.4. Mutual information

Mutual Information (MI) is a criterion commonly used in statistical language modeling of word associations and related applications.[9,10] MI has been used in text classification in Refs.[11-15] and.[7] In our experiments we compute the Mutual Information criterion using:

$$\text{MI}_w = \frac{c_w \times N}{(c_w + c_{\overline{w}}) \times (c_w + \overline{c}_w)}.$$

Mutual Information (MI) has shown some conflicting results. Some report MI as being one of the best performers and others report it as being one of the worst performers. For instance, Refs.[14] and[7] both report that MI performs worse than other methods. However, Ref.[15] report no significant difference in performance between MI and the other feature selection methods.

One weakness of MI is that rare words are emphasized (i.e. they are assigned a high score). In Ref.,[15] rare words which are ranked very high by MI are discarded. This could be a contributing factor to the differences achieved in their performance.[7,15] Joachims[13] agree with this finding and report good performance of MI when rare words are removed from the data.[15]

12.6.5. Information gain

Information Gain (IG) is a frequently employed word scoring metric in machine learning.[16,17] Information gain measures the number of bits of information obtained for category prediction by knowing the presence or absence of a word in a document. In text classification, IG has been employed in Refs.[5,7,14,18,19] In our experiments we compute IG using:

$$\text{IG}_w = -P(c)\log_2 P(c) + P(\overline{c})\log_2 P(\overline{c}) - (P(w) \times (-P(c_w)\log_2 P(c_w)$$
$$- P(\overline{c}_w)\log_2 P(\overline{c}_w))) + (P(\overline{w}) \times (-P(c_{\overline{w}})\log_2 P(c_{\overline{w}}) - P(\overline{c}_{\overline{w}})\log_2 P(\overline{c}_{\overline{w}}))).$$

Information Gain (IG) is another word scoring metric which shows conflicting results. Yang and Pedersen[16,20] report good performance of IG, when compared with other mehods. However, this was not the case in.[14] Mladenic[14] report IG as being one of the worst performers. A reason for this conflict may be due to the fact that different domains and different classification algorithms were used in their experiments.

12.6.6. Odds ratio

Odds Ratio (OR) was proposed in Ref.[20] for selecting words for relevance feedback. It has been used in Refs.[5,15,21,22] for selecting words in text categorization. Odds ratio takes values between zero and infinity. One

('1') is the neutral value and means that there is no difference between the groups compared, close to zero or infinity means a large difference, larger than one means that the relevant set has a larger proportion of documents which contain the word, than the irrelevant set, smaller than one means that the opposite is true. In our experiments we compute the odds ratio using:

$$\mathrm{OR}_w = \frac{c_w \bar{c}_{\bar{w}}}{\bar{c}_w c_{\bar{w}}}.$$

Mladenic,[21] Mladenic et al.[23] and Caropress[5] report that compared to mutual information and information gain odds ratio was the most successful feature selection method. A reason for this may be because Odds Ratio favors words which correlate with c. As a result, words that occur very few times in c but never in $\bar{c}$ will get a relatively high score. Thus, many words that are rare among positive categories will be ranked at the top of the word list.

12.6.7. *Document frequency*

Document Frequency (DF) is the number of documents in which a word (w) occurs. It is a simple and common technique employed in several text categorization problems.[1,8,20,25] In our experiments we compute DF using:

$$\mathrm{DF}_w = c_w + \bar{c}_w.$$

Previous research show that DF gives good performance for feature selection.[5,7] However, DF is considered as an *ad hoc* approach to improve efficiency and it is not considered as a principled criterion for selecting predictive features.

12.6.8. *Fisher criterion*

The Fisher criterion is a classical measure to assess the degree of separation between two classes. We use this measure in text classification to determine the degree of separation of documents which contain word w within the sets c and $\bar{c}$. In our experiments we compute the Fisher criterion using:

$$f_w = \frac{(c_{\mu_w} - \bar{c}_{\mu_w})^2}{(c_{\sigma_w})^2 + (\bar{c}_{\sigma_w})^2}$$

where c_{μ_w} is the mean number of documents which contain the word w and belong to c, σ_w is the standard deviation of documents in c that contain the word w.

12.6.9. BSS/WSS criterion

This feature selection metric has been used in the context of gene selection.[2] As far as we know, this metric has never been employed in the context of text classification. The metric represents the ratio of between sum of squares (BSS) to within sum of squares (WSS) of two groups. In our experiments we compute this metric using:

$$\frac{\text{BSS}(w)}{\text{WSS}(w)} = \frac{\sum_{j=1}^{N} \sum_{C \in \{c,\bar{c}\}} I(d_j = C)(\mu_{C,w} - \mu_w)^2}{\sum_{j=1}^{N} \sum_{C \in \{c,\bar{c}\}} I(d_j = C)(x_{w,j} - \mu_{C,w})^2}$$

where $I(d_j = C) = 1$, if document j belongs to C (where $C \in c, \bar{c}$) and zero otherwise, μ_w is the average occurrence of word w across all documents, $\mu_{c,w}$ denotes the average occurrence of word w across all documents belonging to c. $x_{w,j}$, is the frequency of occurrence of word w in document j.

12.6.10. The GU metric

In statistical analysis, significance testing (z), measures the differences between two proportions. A high z score indicates a significant difference between the two proportions. This is the motivation behind the algorithm. We use the z score to measure the difference in proportions between documents which contain word w and belong to c and those that contain w and belong to $\bar{c}$. The larger the z score the greater the difference in proportions so the word is better as a discriminator of the two classes. We evaluated variations of the raw z score as a feature selection metric. The GU metric uses the following formula:

$$\text{GU}_w = |z| \cdot \frac{c_w \cdot n_{\bar{c}}}{n_c \cdot \bar{c}_w}.$$

Here z is computed as follows:

$$z = \frac{c_w - \bar{c}_w}{\sqrt{p(1-p)\left(\frac{1}{n_c} + \frac{1}{n_{\bar{c}}}\right)}}$$

where

$$p = \frac{\bar{c}_w + c_w}{n_c + n_{\bar{c}}}.$$

12.7. Experimental Setting

In our experiments we chose to use the 20 Newsgroups data set.[24] This data set is widely used as benchmark in text classification experiments. The 20

Newsgroups data set consists of Usenet articles collected from 20 different news groups.[a] For our experiments we train one Naive Bayes classifier for each news group. The task was to learn whether a certain news article should be classified as a member of that news group.

The news groups data was pre-processed before being used in the experiments. Mail headers were removed, only the body and the subject line were retained from each message. The words found in the standard stop-list were also removed and the remaining words were stemmed.

The news articles in each news group were divided into a training set and a test set. Using 80% of the documents to represent the training set and 20% of the documents to represent the test set. For each experiment two sets were formed. Set one, which will be referred to as c from hereafter, contains all news articles of the training set for that news group, set two ($\bar{c}$), is a combination of all remaining news articles from the training sets of all the other news groups. Next each word appearing in c and $\bar{c}$ is assigned a score using one of the word scoring metrics. The words are then sorted according to their individual scores and the top scoring words are selected to represent the feature subset. This feature subset $V = \{w_1, \ldots, w_n\}$, consists of n distinct words. The Naive Bayes probabilistic classifier is then used to decide the most probable class (c or $\bar{c}$) for each news article from the test set, depending on whether the news article contains words from V.

The Naive Bayesian probabilistic classifier is computed using the equation below:

$$c^* = \operatorname{argmax} P(C|d) = \operatorname{argmax} P(C) \prod_{k=1}^{n} P(w_k|C)^{N(w_k, d_C)}$$

where, $C \in \{c, \bar{c}\}$ and $N(w_k, d_C)$ is the number of occurrences of word w_k in news article d_C.

The word probabilities $P(w_k|C)$ are computed using the Laplacian prior (see equation below).

$$P(w_k|C) = \frac{1 + \sum_{d_i \in C} N(w_k, d_i)}{|V| + \sum_{r=1}^{|V|} \sum_{d_i \in C} N(w_k, d_i)}$$

Performance of each word scoring metric was measured by increasing the feature set size by 10 features each time, until a total of 1000 features were selected. Each feature scoring metric was evaluated using the same training and test set.

---

[a] Over a period of time 1000 new articles were taken from each of the news groups and with the exception of a few articles, each article belongs to exactly one news group.

12.8. Empirical Validation

To evaluate the performance of each feature scoring metrics we used the performance measures: precision (p), recall (r), F_1 measure and F_2 measure (see equations below):

$$p = \frac{A}{A+B}$$

$$r = \frac{A}{A+C}$$

$$F_1 = \frac{2pr}{r+p}$$

$$F_2 = \frac{3pr}{r+2p}$$

where A is the number of news articles correctly classified, B is the number of news articles incorrectly classified. C is the number of news articles in the category. The F measures are a combination of both the precision p and recall r metrics. In our experiments we report both the F_1 and F_2 for each experiment.

12.9. Results

The results presented below report the average precision, recall, F_1 and F_2 measures for each category prediction. They are calculated using a set of correctly classified documents. Reported results are averaged over five repetitions using a different training and test set each time.

Figure 12.4 shows the precision versus feature set size. It can be seen that χ^2 and NGL metrics show similar results and they show the best precision scores. Next best is the GU *metric*. The worst precision scores are obtained by Mutual Information. Figure 12.5 shows the recall versus number of features. Here, the best performers are IG and GSS coefficient, next is the GU *metric*. Figure 12.6 shows the F_1 scores versus number of features. Here, the GU *metric* shows the best performance. Figure 12.7 shows the F_2 scores versus number of features, these results show similar results to the F_1 measures.

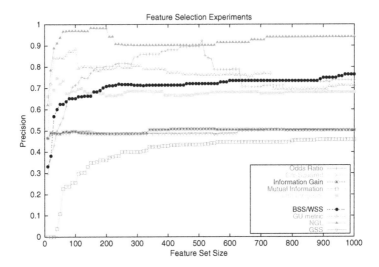

Fig. 12.4. Precision of the feature selection experiments.

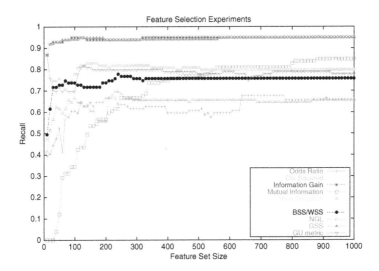

Fig. 12.5. Recall of the feature selection experiments.

12.10. Summary and Conclusions

We have presented a comparative study of existing feature selection methods and some new ones using Lang's 20 Newsgroups dataset, to measure

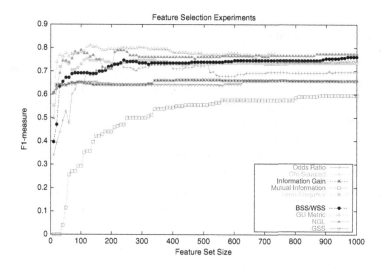

Fig. 12.6. F1 measure of the feature selection experiments.

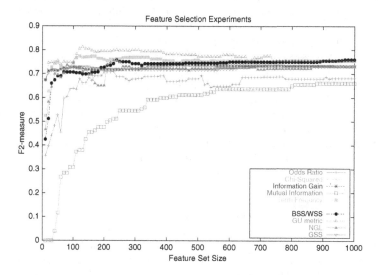

Fig. 12.7. F2 measures of the feature selection experiments.

the performance of each feature scoring methods in text classification.

Our experimental results are not in contradiction with previously reported results of Mladenic,[12] they report that OR had better F_1 scores than IG and MI. This is also what we conclude. The overall worst performer has been obtained by the MI method which is what Mladenic and Yang and Pedersen reported.

In our experiments we do not report a difference in performance between the NGL coefficient and the χ^2. Also, GSS coefficient does not perform better than NGL and χ^2. In our study we can conclude that the best performers using the Naive Bayesian classifier is χ^2, GU *metric*, BSS/WSS, NGL. GSS and IG show similar behavior.

The results which we have obtained from this study are promising. The *GU* metric performs as well as some of the more common feature selection methods such as χ^2 and outperforms some other well known feature selection methods such as *Odds Ratio* and *Information Gain*. Our experimental evaluations are still ongoing. In particular we are continuing experimental evaluations on different domains and using different classifiers.

References

1. G. Uchyigit and K. L. Clark. Hierarchical agglomerative clustering for agent-based dynamic collaborative filtering. In *Fifth International Conference on Intelligent Data Engineering and Automated Learning (IDEAL'04)*, Exeter (August 25-27, 2004). IEEE, Springer-Verlag.
2. G. Uchyigit and K. L. Clark. A multi-agent architecture for dynamic collaborative filtering. In *Proceedings of the 5th International Conference on Enterprise Information Systems*, Angers, France (April 22-26, 2003).
3. G. H. John, R. Kohavi, and K. Pfleger. Irrelevant features and the subset selection problem. In *Proceedings of the Eleventh International Conference on on MachineLearning*, pp. 121–129, (1994).
4. H. Almuallim and T. G. Dietterich. Learning with many irrelevant features. In *Proceedings of the Ninth National Conference on Artificial Intelligence, AAAI press.*, pp. 547–552, San Jose, (1991).
5. M. Caropresso, S. Matwin, and F. Sebastiani. A learner independent evaluation of usefulness of statistical phrases for automated text categroization. In ed. A. Chin, *Text Databases and Document Management: Theory and Practice*, pp. 78 – 102. idea group publishing, (2001).
6. L. Galavotti, F. Sebastiani, and M. Simi. Experiments on the use of feature selection and negative evidence in automated text categorization. In eds. J. L. Borbinha and T. Baker, *Proceedings of ECDL-00, 4th European Conference on Research and Advanced Technology for Digital Libraries*, pp. 59–68, Lisbon, PT, (2000). Springer Verlag, Heidelberg, DE.
7. Y.Yang and J. Pedersen. A comparative study on feature selection in text categorization. In ed. D. H. Fisher, *Proceedings of ICML-97, 14th International Conference on Machine Learning*, pp. 412–420, Nashville, US, (1997). Morgan Kaufmann Publishers, San Francisco, US.
8. H. Ng, W. Goh, and K. Low. Feature selection, perceptron learning, and

a usability case study for text categorization. In *SIGIR '97: Proceedings of the 20th Annual International ACM SIGIR Conference on Research and Development in Information Retrieval, July 27-31, 1997, Philadelphia, PA, USA*, pp. 67–73. ACM, (1997).
9. R. Fano, *Transmission of Information*. (MIT Press, Cambridge, MA, 1961).
10. K. W. Church and P. Hanks. Word association norms, mutual information and leixicography. In *ACL 27*, pp. 76–83, Vancouver Canada, (1998).
11. S. T. Dumais and H. Chen. Hierarchical classification of web content. In *SIGIR'* (August, 2000).
12. S. T. Dumais, J. Platt, D. Heckerman, and M. Sahami. Inductive learning algorithms and representations for text. In *ACM-CIKM*, pp. 148–155 (November, 1998).
13. T. Joachims. A probabilistic analysis of the rocchio algorithm with tfidf for text categorization. In *ICML*, pp. 143–151, (1997).
14. D. Mladenic. *Machine Learning on non-homogeneous, distributed text data*. PhD thesis, University of Ljubljana,Slovenia (October, 1998).
15. M. E. Ruiz and P. Srinivasan, Hierarchical text categorization using neural networks, *Information Retrieval*. **5**(1), 87–118, (2002).
16. T. M. Mitchel, *Machine Learning*. (McGraw-Hill International, 1997).
17. J. R. Quinlan, *C4.5: Programs for Machine Learning*. (Morgan Kaufmann, 1993).
18. G. Forman, An extensive empirical study of feature selection metrics for text classification, *The Journal of Machine Learning Research*. **3** (March, 2003).
19. M. Pazzani and D. Billsus, Learning and revising user profiles: The identification of interesting web sites., *Machine Learning*. **27**, 313–331, (1997).
20. V. Rijsbergen, *Information Retrieval*. (Butterworths, London 2nd edition, 1979).
21. D. Mladenic, J. Brank, M. Grobelnik, and N. Milic-Frayling. Feature selection using linear classifier weights: Interaction with classification models. In ed. ACM, *SIGIR'04*, (2004).
22. Z. Zheng and R. Srihari. Optimally combining positive and negative features for text categorization. In *ICML-KDD'2003 Workshop: Learning from Imbalanced Data Sets II*, Washington, DC (August, 2003).
23. D. Mladenic, J. Brank, M. Grobelnik, and N. Milic-Frayling. Feature selection using linear classifier weights: interaction with classification models. In *Proceedings of the 27th annual international conference on Research and development in information retrieval*, pp. 234 – 241, Sheffield, United Kingdom, (2004).
24. K. Lang. Newsweeder: Learning to filter netnews. In *12th International Conference on Machine Learning*, (1995).

BIOGRAPHIES

Gulden Uchyigit received PhD in computing from Imperial College, London. B.Sc. in computer science and M.Sc. in advanced computing from the University of London. Her research interests are in the area of machine learning, personalization and user modeling. She has authored over 30 papers in refereed books, journals and conferences. She serves on the programme committees of several international conferences and has oranganised and chaired several workshops in the area of personalisation technologies and recommender systems.

Keith Clark is a Professor of Computational Logic at Imperial College. His early research was in the area of the theory of logic programming. His recent research focuses on distributed symbolic programming languages and methodologies for building rational and multi-agent applications.

Subject Index

$\chi 2$, 234, 303, 308

Additive learning, 45
Advisors, 251

Browsing activity evaluation, 47
BSS/WSS coefficient, 303, 308
Buffer rehearsal process, 39

Case-based reasoning, 19
ChangingWorlds, 26
Chunks, 41
Click-distance, 9
Co-occurrences, 45
Collaborative filtering and a spreading activation approach, 173
Collaborative filtering approach: memory-based with Pearson correlation, 171
Collaborative filtering as a social network, 169
Collaborative web search, 17
Community-based search, 18
Compatibility constraints, 252
Conflict sets, 259
Conjunctive queries, 258
Constraints, 251, 253, 254, 256, 257, 259, 265, 272

Consumer buying behavior, 264
Customer properties, 252
CWW, 37

Diagnosis, 270
Digital TV, 191, 192
Document object model, 48
DOM, 48

Electronic Programming Guide, 223
Empirical findings, 263
　explanations, 266
　product comparisons, 267
　pure product lists, 266
　repairs, 268
　willingness to buy, 269
Experiences from projects, 262
Explanations, 259
Explicit feedback, 34, 36

Failed session, 22
Feature selection, 234, 303, 308
Filter constraints, 253, 254
Fisher criterion, 303, 308
Forgetting, 45

Graph-based approaches for recommendation, 168

Graph-based representations of the collaborative filtering space, 173
GSS coefficient, 303, 308

HAL, 45
HAL matrix, 46
Hyperspace analogue language model, 45

I-SPY, 20
Identifying User features from a collaborative filtering dataset, 176
IDF, 42
Implicit feedback, 34, 45
Importance of explanations, 264, 267
Information foraging, 47
Information gain, 303, 308
Information retrieval, 17
Information scent, 47
Inverse document frequency, 42
Inverse vector frequency, 46
IVS, 46

Just-in-Time IR, 35

Knowledge acquisition, 251, 254–256
Koba4MS, 250, 251, 257, 260, 270

Long-term store, 38
LTS, 38

Machine learning, 307
Maximum entropy model, 232
Mobile internet, 5
Mobile portals, 5

Mutual information, 303, 308

Natural language processing, 37, 41
Navigation problem, 8
NGL coefficient, 303, 308
NLP, 37, 41

Odds ratio, 303, 308
Ontology, 192, 195–199, 201–203, 206, 207, 209, 213, 215–217
OWL, 206, 209, 213, 216, 217

Personalization, 3, 4
Personalization strategies
  collaborative filtering, 194–196, 201, 210, 212, 215–217
  content-based methods, 195, 196, 201, 210, 217
  hybrid strategies, 194, 209, 210, 217, 218
Personalization-privacy tradeoff, 25
Probe cues, 40, 44
Probing, 40
Product comparisons, 254
Product properties, 252
Proximal cues, 47
Psychological aspects, 271, 272

Recommendation, 249
  collaborative, 250
  content-based, 250
  knowledge-based, 250, 251
  processes, 251, 255
  user interfaces, 255
Recommender applications
  digital cameras, 260, 270
  financial services, 261

financial support, 262
Recovery process, 39, 42
Renting, 45
Repairs, 258

SAM simulation, 39
SAM theory, 38
Sampling, 42
Sampling process, 38, 42
Search community, 17
Search of associative memory theory, 38
Semantic reasoning, 192, 195, 196
Semantic similarity
   Hierarchical semantic similarity, 207
   hierarchical semantic similarity, 202
   Inferential semantic similarity, 207
   inferential semantic similarity, 202
Semantic web, 192, 195, 217
Short-term store, 38
Sliding window, 45
SNIF-ACT, 37

Sparsity problem, 194, 195, 217
Strength matrix, 39, 41, 43
STS, 38
Successful session, 22

Term frequency, 303
Text classification, 308
text classification, 303
Trust in recommendations, 264, 266

Usage data, 34, 40
User acceptance, 263, 270
User neighborhood, 194, 203–205, 214
User profile, 4, 192
User studies, 263

Vague query problem, 17
Vocabulary gap, 17

WAP
   wireless application protocol, 6
Weighting Schemes in collaborative filtering, 168